AND THE MOON SHALL TURN TO BLOOD

THE PROPHECY TRILOGY - VOLUME I

ANTHONY E. LARSON

ZEDEK BOOKS
P.O. Box 1271 Orem, Utah 84057

Revised L.D.S. Edition, Volume I of *The Prophecy Trilogy*

First Printing, December 1983
Second Printing, May 1993

Zedek Books
Orem, Utah

Printed in the United States of America

Contents

Acknowledgments

There can be little doubt that the research and writings of Dr. Immanuel Velikovsky are fundamental to the concepts considered in this work. His insightful observations were unique among those of his contemporaries. Many before him had sought in vain for the historical affirmation that he found of the people, places and occurrences described in ancient scriptures. In his efforts to correct modern misconceptions of ancient chronology, Dr. Velikovsky discovered important evidence pertaining to events which occurred in the distant past. His theories not only establish the Old Testament as an authentic historical document, but they also give considerable insight into the more cryptic and enigmatic references in the scriptures. Dr. Velikovsky's contribution is therefore immeasurably great.

For their encouragement, support and assistance I gratefully acknowledge my friends Kaye and Michael Jensen. The task of writing this book would have been much greater without their willingness to lend aid and advice.

Last of all I express thanks to my long-suffering family—especially to my parents and to my wife, Sue. They have given the most support and sacrifice to enable me to bring this effort to fruition.

Preface

As you, the reader, embark on the wondrous quest through the past and the future depicted in this book, be prepared to lay aside some of the secular notions you have taken for granted nearly all your life. At the same time, be prepared to have the reality of Biblical events and the words of the Prophets confirmed for you far beyond anything you might have imagined possible. When you finish reading this book, your concept of the Prophets, the past and the future will be changed forever.

Why another book on prophecy?

With all the many books and articles written in the past that attempted to clarify prophecy, some people may be inclined to say that another book on prophecy would be superfluous. Yet with all that has been written about prophecy, most students of the scriptures still struggle to understand it—confirming the existence of a lack of understanding and indicating a need for further investigation. The Book of Revelation in the New Testament is a case in point. Even with the heightened understanding that we enjoy because of the enlightened age we live in, and even with the volumes already written on prophecy, John's vision is still an enigma to most Latter-day Saints.

Perhaps this information gap still exists because past efforts to analyze prophecy failed to give the reader any useful tools to apply to the study of the scriptures. This book, on the other hand, is written specifically to enlighten the reader as to the meaning of the language of prophecy. Equipped with a few simple definitions of prophetic metaphors, and an enhanced understanding of the past, you will be able to discover much of the meaning of prophecy for yourself rather than depending on someone else to interpret it for you.

This book deals with understanding prophecy and prophetic metaphors—not merely collecting and categorizing any number of prophecies. It is not just another book on prophecy. The implications of the novel view of the past and the future considered herein are nearly overwhelming; and if these concepts are indeed correct, they leave little doubt about the future of the children of God and the world in this day and time. This work is an attempt to introduce a new perspective on history and how it relates to prophecy—not an attempt to assemble all prophecy in one volume or an effort to predict the dates of prophesied events. It is hoped that the reader will apply this new perspective to his study of the scriptures and re-examine prophecy *for himself* using these new concepts.

Understanding is everything

It is not a valid argument to say that we needn't understand prophecy because if the Lord had intended us to, he would have given us all interpretations. While this kind of thinking may be correct with respect to some gospel subjects, the sheer preponderance of prophecy in the scriptures argues eloquently against such a conclusion. Why would the Lord have given us so much prophecy if he did not expect us to understand it? While Joseph Smith plainly taught that there were some few images and symbols in scripture that we may not yet understand (because their meaning has not yet been revealed), where is the logic in having so much revelation—whole books in some cases—concerning the last days, written by men of God, only to have the people who live in these last days left ignorant as to its meaning? A more rational analysis suggests that the Lord desires us to study the scriptures in order to learn their meaning—by exercising faith and exerting effort.

All the biblical quotes in this book are taken from the King James version of the Bible. A comparison of these quotes with other Bible translations may be useful to enhance the reader's understanding of the scriptures.

The gospel test

Almost any theory or thesis that touches on religion can be tested by applying gospel principles to it. Latter-day Saints often use principles of the gospel as a means of testing the validity of scientific theories. If a theory conforms to gospel principles, then we accept it as part of all truth which the gospel embraces.

Like most Latter-day Saints, I see things in terms of the truths of the gospel, and whenever I am exposed to new concepts or principles, I test them with the truths of the gospel which I hold dear. The concepts of the past and the future examined in this book are the results of just such a test: applying the truths of the gospel to a new scientific theory. I have found the results fascinating.

It is my sincere hope that you will also find these insights to be enlightening and fascinating, that this book will open new doors of inquiry to your mind, that you will find these fresh insights so compelling and fascinating that you will not be able to ignore their implications and that the concepts discussed in this book greatly enhance your understanding of the scriptures.

"The first angel sounded, and there followed hail and fire mingled with blood, and they were cast upon the earth: . . . and there were voices, and thunderings, and lightnings, and an earthquake."

Revelation 8:5, 7

1

A Scenario for the Future

The hypothetical scenario described in this Chapter is based on the theories examined in subsequent Chapters of this book. Except for this brief fictional depiction, all the other information detailed in this book is derived exclusively from thorough research and is based on scientific analysis.

In this Chapter, you are taken to a not too distant future time when what is here described could very well happen.

I don't know that I can write down how I feel or even begin to describe what's happened in the past two weeks. The world I knew and all the things I thought were firm and immovable have been shaken and changed, until all that's left of the way life used to be is a memory—and even that seems so far away I'm not really sure any more that it was ever real . . .

I remember the day the dust started falling. I had gone out to get into the car to go to work, when I saw the filmy layer of powder all over the windshield. I wiped my hand across the glass and felt the talc-like dust, which turned the ends of my fingers red. I thought it was probably a strange air pollution, so as I got my snow brush out of the car and began sweeping off the windshield, I looked into the distance for a telltale wall of smog. Instead of the brown haze of air pollution, the horizon was a pale red—as though the sun were just rising (even though the sun had been up for over an hour). I looked up at the sky, and that, too, was tinted, somehow, and appeared almost pink.

I brushed most of the dust off the windshield and then tried the wipers. When water hit the dust, it turned a bright red.

Still puzzled, I backed out of the driveway and headed down the street. It seemed that everything was dusted with the powder. I flipped on the radio and heard a bulletin stating that parts of western North America, as far north as Alaska, had reported a fine, rust-colored dust settling out of the atmosphere. In some places the stuff had been quite thick. A report from the Seattle area claimed that after a storm front had moved through that morning, the rivers, lakes and streams had been polluted by a heavy, rust-red rain that had fouled the water supplies and killed fish and other aquatic life. Large metropolitan areas were suddenly faced with the very real danger of being without water.

That evening, we ate dinner in the family room, crowded around the television set as network programming was preempted for reports on what was fast becoming a worldwide phenomenon.

Scientists first speculated that volcanic activity had caused the dust, but no eruptions had been reported anywhere.

Pentagon sources denied rumors that the dust was radioactive and somehow linked with nuclear testing.

Then, in the days that followed, a new type of dust, somewhat like ash, began to appear in the atmosphere. Cases of skin irritation—open sores and even reports of ash-related deaths—began to alarm officials, and we were warned to stay indoors as much as possible.

About that time the real cause of the dust and ash were identified. For months astronomers had been watching a new comet that had entered our solar system far out beyond the farthest planets. It was larger than most comets, even larger than our moon, and had caused quite a stir among the scientists. We'd seen it ourselves. It was much brighter than any star, and it had a tail that stretched halfway across the sky. Astronomers had assured us, at first, that the comet posed no danger to the Earth, but later they admitted that they had been wrong. They now stated that as the giant comet had swung around the sun and started on its way back to where it came from, the gravitational pull of one of the larger planets had altered its course enough to bring the comet perilously close to the Earth—tail first. The rain of dust and ash had come from the debris that made up a portion of the comet's tail.

As the days passed, reports of meteorite showers became more and more frequent. In some places the meteorites were the size of pebbles and pelted the earth like hail in a summer thunderstorm. The damage was devastating. Power and communication systems were knocked out, and the network of communications satellites was virtually destroyed.

The noise last night was terrifying. My family and I huddled in the basement and prayed, while large meteorites rained down on our neighborhood like artillery shells.

Early this morning an earthquake shook our neighborhood, ruptured a nearby dam, and brought the fouled floodwaters within a mile of us. We spent the day doing what we could for neighbors and listening to garbled reports over a shortwave radio. We heard reports of earthquakes and volcanic activity in almost every quarter of the globe. One report even said that the moon had began to glow an ominous red.

We haven't seen the sun for days as smoke and ashen clouds fill the sky—the air gets thicker by the hour. One question keeps forcing itself into my mind: "What is to become of us?"

"For the Lord hath poured out upon you the spirit of deep sleep, and hath closed your eyes: . . . And the vision of all [the Prophets] is become unto you as the words of a book that is sealed. . . ."

Isaiah 29:10, 11

2

A Key to the Past

Those of us who live in this day and time face the very real possibility of experiencing firsthand the events associated with the Savior's Second Coming. Of those events, the Lord has not left us without revelation. The scriptures are replete with prophetic accounts of what is to come, and the Lord has enjoined us to be watchful: ". . . he that feareth me shall be looking forth for the great day of the Lord to come, even for *the signs of the coming of the Son of Man.*" (Doctrine & Covenants 45:39; italics added.)

Nevertheless, prophetic accounts of those events (with which we are commanded to become familiar) seem incomprehensible in many instances. For example, the Savior said of his Second Coming:

> And immediately after the tribulation of those days, the sun shall be darkened, and the moon shall

> not give her light, and the stars shall fall from heaven, and the powers of heaven shall be shaken.
>
> . . . then shall appear the sign of the Son of Man in heaven. . . . (Joseph Smith—Matthew 24:33, 36.)

Many questions

What could cause such things to occur? Understanding, as we do, the true nature of the universe, how can we believe that all the stars will fall from the heavens? Will our sun and all the suns in the universe suddenly go dark? Such a thing seems impossible. And what will be "the sign of the Son of Man in heaven"? We know the Savior will come, but he indicated in these verses from Matthew that a "sign" will precede his coming. What is this marvelous "sign" to be?

These are only a few of the questions that come to mind upon reading two small verses in Matthew. Considering that the scriptures are full of prophecy concerning events to occur before the Savior's Second Coming, such questions multiply a hundred fold in the reader's mind.

We do not understand

But our lack of understanding is not limited to prophecy. Many events recorded in the Old Testament are as puzzling as any of the strange events foretold in prophecy. Perhaps that is because as modern men and women, living (as we believe) in an enlightened age, we frequently labor under the illusion that we are more knowledgeable than our predecessors. *But judging by our failure to understand many of the prophetic utterances of the Prophets, and many of the extravagant events from the past, it seems possible that they may have experienced many things which we have not.* Could it be that the ancients had knowledge of things we are ignorant of? In the Book of Joshua, for example, it is written:

> Then spake Joshua to the Lord in the day when

> the Lord delivered up the Amorites before the children of Israel, and he said in the sight of Israel, Sun, stand thou still upon Gibeon; and thou, Moon, in the valley of Ajalon.
>
> So the sun stood still in the midst of heaven, and hasted not to go down a whole day. (Joshua 10: 12, 13.)

Current scientific logic dictates that such a thing could never happen. The orderly movement of the heavens and the Earth simply could not be interrupted in the manner described by Joshua. What, then, did Joshua and the Israelites experience? Why are such strange and extravagant statements found in scripture? Are we expected to believe that such things really happened?

The Exodus

Another example of such a difficulty is the account written by Moses. Nothing in our modern experience serves fully to explain the cause or causes of the miracles associated with the Exodus:

> . . . and all the waters that were in the river were turned to blood. (Exodus 7:20.)
>
> So there was hail, and fire mingled with the hail, very grievous. . . . (Exodus 9:24.)
>
> And the locusts went up over all the land of Egypt . . . For they covered the face of the whole earth, so that the land was darkened. (Exodus 10: 14, 15.)
>
> And it came to pass, that at midnight the Lord smote all the firstborn in the land of Egypt. . . . (Exodus 12:29.)
>
> And the Lord went before them by day in a

> pillar of a cloud, to lead them the way; and by night, in a pillar of fire, to give them light; to go by day and night: . . . (Exodus 13:21.)
>
> . . . and the waters were divided . . . . and the waters were a wall unto them on their right hand, and on their left. (Exodus 14:21, 22.)
>
> . . . they said one to another, It is manna: for they wist not what it was. And Moses said unto them, This is the bread which the Lord hath given you to eat. (Exodus 16:15.)

Is it possible that a better understanding of the account written by Moses would give us insights into the world of the Prophets—and the similes and metaphors of prophecy itself? Without losing sight of our goal of understanding prophecy, let us turn momentarily to the account of the Exodus.

Moses wrote of bloodied water, of a pillar of smoke and fire, of walls of water, and of manna from heaven. Perhaps his words served as an apt description of what he saw, but they seem to offer little assistance in determining what those phenomena really were. It appears that Moses desired to impress upon his readers' minds the power and greatness of God, rather than explaining the natural causes of the miracles associated with the Exodus.

Miracles are comprehensible

In seeking to understand the *means* whereby God may have performed some of the miracles recorded in scripture, we do not rob God of his holiness or power. His omnipotence is not dependent upon human ignorance. Elder James E. Talmage wrote that "miracles cannot be in contravention of natural law, but are wrought through the operation of laws not universally or commonly recognized." (*Jesus the Christ,* p. 148.) Just as God calls certain men to be his agents, to act in his name and do his bidding here upon the Earth, so also he may direct the elements of nature to do his bidding—acting as his agents. If so, then miracles are not metaphysical events; rather, they may be the results of the natural elements obeying their Creator's commands. Therefore, the miracle of "the pillar of fire and smoke" may have been a manifestation of the power of God through natural causes, as indeed all the miracles of the Exodus. It appears that if we were sufficiently knowledgeable in these things, we would understand the causes of such things.

A comparison—past and future

There may be great value in attempting to understand the true nature of the Exodus miracles, for we find that many of the signs and phenomena discussed in ancient and modern prophetic writings are very similar to the similes and metaphors employed by Moses. For example, compare the following excerpts from the Books of Exodus and Revelation. Notice the striking similarity in the descriptions of past and future events:

The Book of Exodus (history)	**The Book of Revelation (prophecy)**
. . . and all the waters that were in the river were turned to blood. And the fish that was in the river died; and the river stank, and the Egyptians could not drink of the water of the river; and there was blood throughout all the land of Egypt.	And the second angel poured out his vial upon the sea; and it became as the blood of a dead man: and every living soul died in the sea. And the third angel poured out his vial upon the rivers and fountains of water; and they became blood.

and there followed hail and fire mingled with blood, and they were cast upon the earth. . . .	. . . and the Lord sent thunder and hail, and the fire ran along the ground. . . .
And the hail smote throughout all the land of Egypt all that was in the field, both man and beast; and the hail smote every herb of the field, and brake every tree of the field.	And there fell upon men a great hail out of heaven, every stone about the weight of a talent: and men blasphemed God because of the plague of the hail; for the plague thereof was exceeding great.
And it shall become small dust in all the land of Egypt, and shall be a boil breaking forth with blains upon man, and upon beast, throughout all the land of Egypt.	. . . and there fell a noisome and grievous sore upon the men . . . and they gnawed their tongues for pain, and blasphemed the God of heaven because of their pains and their sores. . . .
And all the people saw the thunderings, and the lightnings . . . and the whole mount quaked greatly.	. . . and there were voices, thunderings, and lightnings, and an earthquake.
. . . and the voice of the trumpet exceeding loud; so that all the people that was in the camp trembled.	And the seven angels which hold the seven trumpets prepared themselves to sound. The first angel sounded. . . .

An understanding of the miracles of the Exodus would aid us in understanding the Book of Revelation specifically, and all prophecy in general. But because the Book of Exodus does not explain the cause or causes of the miracles which Moses described, it appears as though we lack the information necessary to pursue such an inquiry.

Faced with this dilemma, our investigation must turn to other sources. Were there other witnesses to the events of the Exodus that could corroborate the Old Testament account and provide additional information? Did the people of other nations see the incident? If so, what did they see, or think they saw?

Other witnesses

The Egyptian nation certainly would be the most likely candidate of investigation for us—but which Egyptians? Moses did not indicate the year in which the Exodus occurred, and modern historians have been unable to agree upon a date for that event. In fact, the lack of correlation between Egyptian and Hebrew history has proven to be a major obstacle. Until recently, scholars believed that much of Hebrew history (including the account of the Exodus) was merely a collection of fables rather than accurate history. They had been unable to find any ancient Egyptian record which corroborated the story of the Exodus, and it appeared that no one else on Earth had witnessed the events Moses described.

However, an excellent clue to the dilemma of the missing Egyptian witnesses was supplied by the scholar and historian, Immanuel Velikovsky. In a treatise entitled *Theses for the Reconstruction of Ancient History* (1945), Dr. Velikovsky advanced a revised chronology of ancient history. It was his thesis that the Egyptian chronology universally acclaimed by scholars as an accurate timetable of ancient events was, in fact, a seriously flawed chronology. His thesis contained 284 basic points substantiating his reconstruction. A subsequent book entitled *Ages In Chaos* (published in 1952) gave an even more detailed explanation—supported by exhaustive documentation—covering the period of time from the Exodus to the reign of Pharaoh Ahknaton. (See Appendix A for a biographical sketch of Dr. Velikovsky.)

What was lost is found

Impressed with what appeared to be an eyewitness account (in Exodus) of a tremendous natural disaster, Velikovsky began looking for Egyptian records that might substantiate the Biblical record. He discovered a translation by a preeminent Egyptologist, Sir Alan Gardner, of a document entitled *Admonitions of an Egyptian Sage,* which Gardner had ascribed to a period immediately following the end of the Egyptian Middle Kingdom (approximately 1500 B.C.). The papyrus was authored by an Egyptian scribe named Ipuwer. Gardner had not seen the connection between the *Papyrus Ipuwer* (as Velikovsky called it) and the Old Testament account of the Exodus. But the correlation was obvious to Velikovsky. In a book describing Velikovsky's discovery of the Papyrus Ipuwer, Dr. C. J. Ransom, a physicist, noted:

> The description of events by Ipuwer was strikingly similar to the narrative of the Exodus. When he made the translation, Gardner did not think that the papyrus discussed the Exodus, so the similarities of the descriptions of events could not be attributed to biased translation. Still, the descriptions are so similar in some cases that, even for one only vaguely familiar with the Hebrew version, it is difficult to distinguish between portions of the Hebrew and Egyptian records. (*The Age of Velikovsky,* p. 15. See Appendix D for a comparison of the *Papyrus Ipuwer* and Exodus texts.)

Thus, Egyptian records had been silent concerning the Exodus, not because the Exodus had not occurred—as many skeptics had supposed—nor because the Egyptians had failed to record it. Instead, faulty chronology had misled scholars in their search for corroborating testimony. Prevailing opinion dictated that the Exodus probably occurred during the Late Egyptian Kingdom, around 900 B.C. But the *Papyrus Ipuwer* dated from some 500-600 years earlier, at the end of the Egyptian Middle Kingdom. That meant that any information about the Exodus would most likely be found in documents that dated from the same time as the *Papyrus Ipuwer.*

Having found one source of corroborating testimony, Velikovsky began an intensive search of documents from other ancient cultures—contemporaries to the Egyptians and the Israelites—who lived in and around the Middle East at the time of the Exodus. He was looking for accounts that dated from the period shortly after the end of the Egyptian Middle Kingdom. And instead of looking for accounts of metaphysical events or miracles, he was looking for descriptions of a great natural catastrophe. That fresh perspective enabled him to recognize similarities, where others had seen none. It was a key to unlocking the past.

In the writings of each culture he examined (some of them far distant from the Middle East), Velikovsky found more information concerning the phenomena of the Exodus. He became convinced that he was dealing with an event of worldwide proportions. A brief summary of the information he found is astounding:

Egyptians: saw water turn to blood, followed by great signs in the heavens.

Babylonians: saw not just the water, but the whole world turn red; saw a pillar of fire and felt an incredible earthquake that leveled their cities.

Persians: felt the Earth shake, and saw fire rain from the sky.

Meso-Americans: saw fire and stones fall from the sky, and all the world became dark.

Chinese: it became very dark, and the ocean rose up and spilled over the whole continent.

It became evident that the phenomena associated with the Exodus had not been restricted to a local event in Egypt and the Sinai Peninsula. They had been part of a *global cataclysm.*

Much of Velikovsky's work dealt with the cause of those phenomena, and his discoveries shed new light not only upon the Exodus but also upon other Old Testament writings in a way that renders those accounts much more intelligible.

A matter of time

Velikovsky's reconstruction of history has solved what was a most difficult problem. Historians, classicists and scholars had concluded that Hebrew history—the Old Testament—was badly flawed because it did not match the chronology established by Egyptian history. Velikovsky pointed out that just the opposite was true: modern Egyptian chronology (the accepted standard for all ancient chronology) was in error—in some cases by as much as 500 years or more! It was apparent to Velikovsky that the flawed Egyptian chronology had been responsible for distorting Greek history, creating vast periods of time or "dark ages" which could not be accounted for. And because the accepted dating was perverse, the results of archaeological finds sometimes indicated that an entire culture had flourished in an area hundreds of years before it was thought to have existed in the region. This had caused major confusion and contradiction in the study of ancient cultures.

For example, during the excavation of Ugarit (the ancient city of Ras Shamra), conventional chronology indicated that the site belonged to the Thirteenth or Fourteenth Century, B.C. Yet, after those dates had been established, some tablets were found which contained Hebrew. The language was unexpected at that site because it was thought that the Israelites had conquered the city at a much later date. Because of the conventional dating, archaeologists had expected to find evidence of crude, barbarian Canaanites. Instead, they found the tablets—evidence of a rather sophisticated Hebrew culture. How could the Canaanites have been using such writing (including its own special idioms and customs) five to six hundred years before the Hebrews were thought to have entered Canaan?

It turned out that conventional chronology was to blame. Velikovsky's revised dating placed the Israelites in Ugarit at the very time period indicated by the evidence found at the site. They had driven out the Canaanites following the Exodus at a much earlier date than conventional chronology had indicated.

A multitude of such confusing and contradictory examples in the dating of ancient chronology can be corrected by applying Velikovsky's reconstruction. As another instance, this same kind of correction serves to validate the Old Testament account of the Battle of Jericho, correctly correlating with the archaeological evidence from that site.

Geologic change

Velikovsky's work also evaluated two theories of how change takes place on the Earth (as well as in the solar system and the universe).

The first theory is termed *uniformitarianism,* which holds that all geological phenomena are the result of existing forces operating uniformly ever since the earth took shape. This is the prevailing belief of the scientific community. It dictates that the gradual, almost imperceptible changes occurring around us today are the result of the same forces that over eons created lofty peaks, deep canyons, prairies, islands and great continents. In addition, this theory of uniformity extends to all the natural sciences, including astronomy and biology.

The second theory is that of *catastrophism,* which maintains that geologic change results from sudden, dramatic upheavals—implying that when change comes, it is cataclysmic in nature.

Until the time of Darwin in the middle 1800s, scientists and scholars felt that Earth's geologic record clearly supported the idea of catastrophic change. Yet with the dawn of the sciences as we know them today, the belief in a uniform rate of change—gradual but inexorable—became prevalent.

Thus, in 1950, when Velikovsky offered his theory of *recent* planetary disturbances, implying *cataclysmic* rather than uniform change, he was roundly condemned as a heretic by the prevailing scientific community.

The scriptures and catastrophism

Interestingly enough, the scriptures have something to say about the way change takes place on Earth and in the heavens.

The Apostle Peter seems to have had foreknowledge of the misconceptions that would exist among men living in the last days:

> Knowing this first, that there shall come in the last days scoffers, walking after their own lusts, and saying, Where is the promise of his coming? for since the fathers fell asleep, *all things continue as they were from the beginning of the creation.* (II Peter 3:3-4; italics added.)

It would seem that Peter understood that people in the last days would profess uniformitarianism. He continued by pointing out their error:

> For this they willingly are ignorant of, that by the word of God the heavens were of old, and the earth standing out of the water and in the water:
>
> Whereby the world that then was, being overflowed with water, perished:
>
> But the heavens and the earth, which are now, by the same word are kept in store, reserved unto fire against the day of judgment and perdition of ungodly men. (II Peter 3:5-7.)

Thus Peter condemned those who hold such views as being "willingly ignorant" that the heavens and the Earth of old have undergone radical changes, and that the heavens and the Earth we now know are not the same as those which existed in ancient times, for "the world that then was . . . perished." Not only did cataclysmic changes occur in the past, but such changes will again occur, brought on by cataclysmic conditions:

> But the day of the Lord will come as a thief in the night; in the which *the heavens shall pass away with a great noise, and the elements shall melt with fervent heat, the earth also and the works that are therein shall be burned up.* (II Peter 3:10; italics added.)

It appears from these statements by Peter that from time to time during the Earth's existence, great catastrophic changes have occurred. Peter cited the time of the Flood as a time of great changes in the heavens and the Earth. This seems to be a theme running through the whole of scripture.

During prolonged periods of peace and stability in the solar system, the elements do operate in a uniform manner.

Living, as we do, in such a period, it is hard to imagine anything shaping our world other than the familiar forces which we see operating daily around us. Yet if we are to believe the sacred records bequeathed to us, the ultimate forces that truly shape our world are far greater than we may imagine.

The past and prophecy

Peter declared that the world had been thus altered in the past, and that in the last days the same forces would again alter the heavens and the Earth.

Thus, if the forces which changed the Earth in times past are to change the planet again, the records of ancient cataclysm are a key to the future. Understanding the causes underlying the similes and metaphors used in such ancient accounts may both unlock the past and provide an explanation of the same similes and metaphors as they are used to describe future events.

What, then, are the forces that caused past catastrophes such as the Flood? What is it that unleashed the calamities that Peter spoke of as changing the heavens and the Earth? What agents in nature caused the impressive signs of the Exodus?

"Hear this, ye old men, and give ear, all ye inhabitants of the land. Hath this been in your days, or even in the days of your fathers? Tell ye your children of it, and let your children tell their children, and their children another generation."

Joel 1:2, 3

3

The Signs of the Exodus

As a proponent of catastrophism, Velikovsky gave a cosmological explanation for many of the phenomena spoken of in sacred as well as profane history. It was his thesis that the world was nearly destroyed in the past, within the memory of mankind, when celestial bodies very nearly collided with the Earth.

Accounts of these near collisions exist today in the written and oral traditions of the peoples of the Earth. The Old Testament is one among many such ancient records which are filled with eyewitness accounts of global destructions. Furthermore, the natural sciences hold much evidence bearing upon an explanation of the changes the Earth has undergone since the days of the Patriarchs.

The following paragraphs summarize Velikovsky's explanation of what caused the phenomena associated with one of ancient history's most significant cataclysms, the Exodus:

Aware of a link between the circuit of heavenly bodies and the catastrophic ruin of previous generations, the ancients ceaselessly watched the planetary movements. Their traditions recalled that when old epochs dissolved, the new 'Age,' or 'Sun,' was marked by different celestial paths. Astronomers and seers diligently watched for any change which might augur approaching destruction and the end of an age.

Prior to the second millennium B.C., ancient Hindu records spoke of four visible planets, excluding Venus. Babylonians, meticulous in their observations, likewise failed to report Venus.

But long before 1500 B.C., Jupiter, for centuries chief among the deities, shattered the serenity of the skies. A brilliant, fiery object, expelled from that planet, entered upon a long elliptical orbit around the sun. The feared god Jupiter had given birth to the comet and protoplanet Venus.

Terrified, men watched the 'bright torch of heaven' as it traversed its elongated orbit, menacing the Earth. Venus, a Chinese astronomical text recalls, spanned the heavens, rivaling the sun in brightness. 'The brilliant light of Venus,' records an ancient rabbinical source, 'blazes from one end of the cosmos to the other.'

The fears of the star watchers were justified. As Venus arched away from one perihelion passage during the middle of the second millennium B.C. (ca. 1450), the Earth approached this intruder, entering first the outer reaches of its cometary tail. (*Velikovsky Reconsidered,* pp. 6-7; see also Appendix B for a summary of events as postulated by Velikovsky.)

The Exodus and its miracles provide a typical chronicle of the natural plagues and disasters accompanying a near collision of the Earth with a large comet/planet very nearly the size of the Earth. Most of the plagues, destructions and miracles which we read about in conjunction with the Exodus are simply physical manifestations resulting from two worlds in near collision—one an inhabited planet, the other a planet-sized comet.

Water turned to blood

As the comet approached the Earth tail first, what initially entered Earth's atmosphere was the finer material of the fringes of the comet's tail. This fine dust was water soluble, and it turned water red. Hence the scriptures declare, "the waters that were in the river were turned to blood." (Exodus 7:20.) The fouled water killed the fish and was apparently unhealthy for human consumption "for they could not drink of the water of the river." (Exodus 7:24.) In addition, human as well as animal skin was irritated by the dust, causing boils and blains. This condition would have been aggravated by the lack of pure water in which to bathe.

It is thought that the fine dust may have been some form of ferric oxide. Several scientists (Dollfus, Moroz, Binder, Cruikshank, and Fish)—supported by evidence from America's Viking space lander—have confirmed that the surface of Mars is covered with a ferric oxide known as limonite. According to Velikovsky, as the comet Venus traveled through the solar system, it had encountered Mars as well as the Earth. Thus, the limonite on Mars may have been deposited by Venus, and if so, Venus undoubtedly also dusted the Earth liberally with the same material and turned the Earth's surface red for a time.

Plague of vermin

The heat generated by such a close planetary encounter would cause insects and other vermin to propagate at an accelerated rate. A good example of this may be found during a khamsin or sirocco (hot, electrically charged winds that blow in the desert). The area around desert villages during such winds literally teems with vermin.

The heating would be worldwide, but only the vermin indigenous to a particular area would abound in that area. The fertile Nile valley, for example, would provide an ideal habitat for many types of vermin. Hence, at the time of the Exodus, hoards of vermin suddenly appeared and ravished the countryside. Because gestation periods vary according to species, the plague of flies would be seen to precede the plague of frogs, the plague of locusts would be seen to precede the plague of venomous serpents, and so on. It would have seemed to the inhabitants of Egypt that wave after wave of vermin descended upon them.

Hail of stones

As the comet moved closer, the Earth found itself deeper and deeper within the tail, and the material entering the atmosphere became progressively more coarse. First it was a fine red dust, then "small dust" fell (Exodus 9:9), and finally small stones. Of course the small stones fell as meteorites—hot, fiery rains that crushed and burned the crops.

Because the Old Testament translators chose the word "hail," we mistakenly envision a great icy hailstorm akin to those we encounter today. But Velikovsky pointed out that the word "barad" in the original Hebrew text would have been more correctly translated "meteorites." He noted that according to Midrashic and Talmudic sources, the hail of the Exodus was hot. This description more nearly fits meteorites than it does a hail of ice.

As larger and larger meteorites rained down, the Egyptians became increasingly more terrified by the great explosion-like noises which the meteorites made. (Exodus 9:28.)

Consuming fire or burning liquid

In addition to the meteorites, a sticky, burning fluid apparently also fell to earth. Some of it ignited, consuming everything it touched. Some of it collected in pools or fissures, or floated on the surface of the water. The substance came from the atmosphere of the comet when it passed close enough to combine with Earth's atmosphere. Evidence from ancient cultures all over the world (Popol Vuh, Annals of Cuauhtitlan, Book of Exodus, Midrashim, etc.) establishes that a sticky, burning fluid was associated with the encounter. Velikovsky maintained that this fluid was in the nature of naphtha, a highly flammable hydrocarbon compound.

In addition to the volatile liquid, voluminous quantities of flammable gases entered Earth's atmosphere and, mixing with the oxygen, exploded when ignited. Some explosions occurred high in the atmosphere while others occurred at ground level, creating a tremendous holocaust in some places.

This phenomenon implies that the atmosphere of the planet Venus contained large amounts of flammable hydrocarbons, an issue still hotly debated in scientific circles. However, recent discoveries indicate that some planets and moons in the solar system possess high concentrations of methane and other, more complex, hydrocarbon compounds in their atmospheres.

Earthquake

As the distance between the Earth and the comet continued to decrease, each was seized in the gravitational grip of the other while the interplay of tremendous electromagnetic forces altered the Earth's rotation. The Earth's forces of inertia and momentum adjusted to these changes, and its crust reacted violently. A massive earthquake convulsed the entire planet. Whole cities collapsed in some places; others suffered less, but all were affected by the temblor. In addition, the comet's gravitational pull caused the Earth's surface to bulge, rupturing the crust and raising mountain chains in some areas while opening vast fissures in others. This resulted in nearly constant seismic activity throughout the duration of the encounter and for a considerable time thereafter.

In connection with the crustal upheaval, Velikovsky pointed out a translation problem that has created a misunderstanding of the passage concerning the death of the "firstborn" in Egypt. He indicated that the original text of the Old Testament has been badly corrupted. Rather than "firstborn," the text would have been more correctly translated "high-born" or "elect." The high-born Egyptians who lived in houses of rock or hewn stone would have been in more danger of being destroyed by a large earthquake than the Israelites who lived in huts of mud and reeds. So it appeared to the Israelites that the destroying angel had passed over them to destroy the high-born Egyptians, as the Lord had promised.

Darkness

Two effects of this close encounter between comet and planet caused great darkness. The dust and debris composing the tail of the comet were swept into the Earth's atmosphere, blotting out the sun. In addition, volcanic ash—created by the volcanic activity which the great earthquakes had initiated—was ejected into the Earth's atmosphere and added to the already gloomy aspect of the heavens. All this served to make the air in some places so thick that people could scarcely breathe. Fires could not be lit outdoors and nothing could be seen.

Contrary to the impression conveyed by the Biblical account (which implies that only the Egyptians suffered death in these plagues) rabbinical sources declare that 49 out of every 50 Israelites perished in this plague of darkness. Although that number seems inordinately high and thus may be an exaggeration, it serves to emphasize the reality of those plagues as indiscriminate natural events. Imprudent Israelites suffered equally with the Egyptians—as did all Earth's inhabitants who were unprepared for those destructions.

Other accounts which date from the same period as the Exodus, and describe identical destructions, also tell of prolonged days or nights. Velikovsky ascribed this to changes in the Earth's rotation or to changes in the tilt of the Earth's axis. Either or both of these conditions would have left large portions of the Earth's surface exposed to the sun for a prolonged period, causing the inhabitants to declare that the land was "aflame." Peoples on the dark side of the Earth bemoaned the loss of the sun, declaring that it had been destroyed or stolen away.

Heated bodies

The surface of the Earth became unusually hot in some places because of three factors. First, heating by induction (caused by the interaction of the tremendous electromagnetic fields each body possessed) simply caused the ground in some areas to grow hot underfoot. In some places the seas boiled. Not even the moon escaped the holocaust, for without an atmosphere to dissipate the heat, it simply became incandescent—and glowed red. Second, as we have seen, when the Earth's rotation slowed or stopped, portions of the globe were subjected to prolonged exposure to the sun. Third, the gravitational influence of the comet pulled the interior of the Earth, its molten core, toward the crust. This too caused the Earth's surface to become hot.

Pillar of fire and smoke

As the comet approached the point of closest passage, its head and tail appeared together through the overcast sky as a pillar or column of light. To the Israelites this was a sign, or an angel, from God. At first it stood before them as they fled Egypt. Then, as Pharaoh's army approached, it moved from before them to behind them. (Exodus 14:19.)

It may be that the Earth's rotation, which had slowed or stopped earlier, resumed in the opposite direction. Perhaps the poles reversed themselves, making the pillar appear to be 180 degrees removed from its previous position. Equally possible, the comet may have passed the Earth at that moment and moved from one horizon to the other.

In any case, the head and tail of the comet stood perpendicular to the horizon, and appeared as a pillar through the darkened sky—a pillar of smoke by day and a pillar of fire by night.

Mighty wind

Displaced and agitated by the abnormal gyrations of the Earth and the gravitational pull of the approaching comet, the atmosphere of the Earth created hurricanes of incredible strength and velocity.

Walls of water

Among the most compelling and enigmatic signs of the Exodus was the parting of the sea which allowed the Israelites to escape to safety. In the Velikovsky scenario, these walls of water were not as close as Cecil B. DeMille portrayed them in his film epoch, but they were far more awesome. Tides created by the gravitational pull of the comet would have been thousands of feet high. The Israelites would have seen them standing far above and beyond the horizon. (Exodus

15:8.)

A concurrent phenomenon would have been the uncovering of huge tracts of seabed and ocean floor. Just as the Israelites saw the tides rising beyond the horizon, they also saw the seabed before them uncovered and dried by the wind.

The spark

As the last of the Israelites struggled to reach the opposite side of the seabed, a giant spark—similar to a lightning bolt seen in a thunderstorm, yet many times greater—apparently arced between the Earth and the comet. This served to lessen the attraction of the comet on the waters of the Earth. Released from the comet's grip, the water rushed to settle into its former place and engulfed the hapless Egyptians who had pursued the fleeing Israelites. Apparently there were many such electrical discharges during the Exodus, especially during the time of the law-giving at Mt. Sinai.

Food from God

No matter what it was called—manna, ambrosia, heavenly bread, food of the morning dew, honey from the clouds, amrite, or great dew—it was what the people of the world ate for years after the great destruction. If not for the honey-flavored, bread-like stuff which precipitated from the Earth's super-saturated atmosphere, man and animal alike would have perished.

Considering this phenomenon, Velikovsky advanced the theory that the Earth's atmosphere had become charged with hydrocarbon molecules. The action of microbes, abundant in the Earth's atmosphere, converted the hydrocarbons to edible carbohydrates which precipitated out of the super-saturated morning air. In light of this idea, it is interesting to note that since 1971, edible protein has been manufactured from petroleum in Great Britain and France and mixed with other ingredients to make a highly nutritious animal feed. The technology exists for producing edible protein and fats directly from petroleum. (See *The Age of Velikovsky,* pp. 83, 84.) Such technology demonstrates the likelihood of the accuracy of Velikovsky's theory.

Worthy of note is the fact that as the precipitate—the Israelites' manna—fell on water and dissolved, an unusual thing happened. The water turned a creamy white color and became sweet to the taste. Hence the phrase "a land flowing with milk and honey" was not so much poetic as a statement of fact.

Theophany

As the children of Israel encamped at the foot of Mt. Sinai and witnessed the law-giving, the comet, according to Velikovsky, once again approached the Earth. The arc of the Earth's orbit had intersected the course of the comet at two points some distance apart in space. At this time, there was smoke, fire, vapor, thunder and lightning. Of course, we have seen what each of these signs entails.

Unlike the first approach, however, the second was accompanied by voices. The Bible tells us that the children of Israel heard the voice of the Lord, but apparently they were not alone in doing so. In China, the emperor Fang-heun took the name *Yahou.* The Indians of Puget Sound have a sacred word of exclamation: *Yahu.* Mexican Indians say *Yao* or *Yaotl.* In Indonesia there is an oath: *Ju ju huwe* (the "j" is pronounced with a "y" sound). In other cultures the word is variously *Jo* or *Jove* (the "v" is pronounced with a "w" sound), *Iao, Yo,* etc. The children of Israel heard *Yahweh* (written *jhwh*). The similarity of all these words from many cultures indicates that apparently the voice of the Lord was heard by more than just the Israelites alone. (See *Worlds in Collision,* pp. 112, 113.)

Then there was the sound of a trumpet. Velikovsky theorized that this sound may well have been the result of the two charged bodies passing one another in space, the magnetic oscillations between them becoming audible in the atmosphere of Earth. As the distance between them changed, so did the pitch of the tone the oscillations produced. Even the volume increased and decreased. Thus, all the world heard the roar of the law-giving while Moses was on Mt. Sinai during the time of great upheaval.

Battle in the sky

As the comet approached the Earth the second time, a display of cosmic proportions arrayed itself before all humankind. The comet was then close enough that its brightness broke through the clouds and its head and tail appeared to be separate entities—the head, a light-god; the tail, a snake or serpent.

Bolts of lightning leapt between the head and the tail of the comet, causing the tail to move or "writhe" as a snake. The god was "doing battle" with the serpent, and the conflict was accompanied by great meteorite showers falling to Earth.

As the comet moved away, the pageant disappeared in the clouds which enshrouded the Earth. But the crashing of the meteorites and the groaning of the ground convinced many of the Earth's inhabitants that the god of light had indeed defeated the monstrous serpent which had fled underground beneath the Earth and lay there moaning.

This battle in the sky is almost a universal theme in the literature of cultures around the world. It so impressed its witnesses that it was recorded by ancient peoples in their myths, monuments, records and rituals.

Collapsed sky

After their second orbital intersection, the two bodies moved away from one another in space. As they did so, the effects of the comet on the Earth began to diminish and then subside. After the major darkness lifted, however, the sky continued to be blackened by dust and ash. The world was shrouded by low, slow-drifting clouds. According to Midrashic sources, the Israelites wandered for years without seeing the sun. Although they were able to distinguish day and night by the pale glow of the clouds during the day, the Israelites saw the sun and the moon clearly for the first time since leaving Egypt only at the end of their wandering.

Prophetic metaphors

These, then, are the phenomena associated with a near collision of the Earth with a planet-sized comet:

Water turns to blood—a fine, pigmented dust of extraterrestrial origin settles on land and sea.

Plague of vermin—the heating of the Earth's surface causes rapid propagation of some insects and reptiles, and some areas teem with vermin.

Hail—meteorites from space crash to Earth.

Fire—burning naphtha and flammable gases fall to Earth.

Darkness—dust and debris from space combine with volcanic ash to blacken the atmosphere.

Long day or night—Earth's rotation is disturbed or halted.

Earthquake—a global consequence of the gravitational interaction of the two orbs involved, shakes and sunders the planet.

Moon turns to blood, seas boil—heat generated in the encounter raises the temperature of the Earth's land and sea and turns the moon's surface red as a result of its incandescence.

Hurricane, mighty winds—tidal effects of the comet on the Earth's atmosphere create powerful wind actions.

Walls of water, oceans heaped up—another tidal effect of the comet disrupts the Earth's oceans.

Terrible lightnings, lightning bolts—electric discharges take place between the two bodies as they pass each other, as well as between the head and tail of the comet.

Pillar of fire and smoke—the head and tail of the comet are described as seen through a darkened or cloud-filled sky.

God of light—the brightness of the head of the comet is so described.

Leviathan, serpent, snake, dragon—the appearance of the writhing, changing tail of the comet is thus described.

Manna, ambrosia, etc.—edible carbohydrates are produced in the Earth's atmosphere by microbial action on petroleum substances from the comet's atmosphere; it then precipitates with the morning dew.

Land flowing with milk and honey—the effect of manna is described when dissolved in water, turning it a creamy white color and making it sweet to the taste.

Voice of God—sound is described that accompanied the groanings from within the Earth, heard by all men.

Trumpets—sound is created by electromagnetic oscillations between the two bodies and rendered audible in the Earth's atmosphere.

Collapsed sky—low-hanging clouds shroud the Earth for a prolonged period of time.

Other planetary encounters

The near collision at the time of the Exodus was not unique in Earth's history. It was only the beginning of a series of such encounters. Velikovsky explained that approximately 50 years after the Exodus, as Joshua pursued the Canaanites in his effort to take possession of the Promised Land, Venus again drew near. The sun and the moon stood still above the Earth as rotation slowed and stopped. Fire and stones fell from the sky, and the world shook again. Although this encounter was not as devastating as the earlier encounter of the Exodus, the Earth reeled under the tremendously destructive power of Venus.

For centuries after the Exodus, all of the Earth's inhabitants anxiously watched Venus, fearing renewed cataclysms. Inasmuch as the errant protoplanet had traversed the orbits of several other planets, it was inevitable that other planetary encounters would ensue. However, it was Mars, not the Earth, that suffered most from repeated encounters with Venus. Stargazers watched breathlessly as the "planet-gods" Venus and Mars did celestial battle. Being much larger than Mars, Venus emerged the victor, dislodging Mars from its orbit and sending it careening across the Earth's orbit while Venus settled into a stable orbit about the Sun.

Thus Mars, once a pacific planet, became an agent of destruction. The first Earth-Mars encounter caused the calamities associated with the founding of Rome, circa 747 B.C. Its destructive power, which altered the Earth's orbit, won for Mars the title of "Lord of War." Lacking the destructive power of Venus, Mars did somewhat less damage to Earth because of its smaller size, reduced gravitational impact and lack of flammable atmosphere. Nevertheless, Mars became feared worldwide as "the god of battle" and "the destroyer of armies and nations."

On at least two additional occasions—once in 721 B.C. and again in 687 B.C.—Mars shook the Earth as it passed, repeatedly changing the Earth's orbit, tilting the Earth's axis and throwing its elements into chaos.

The last Earth-Mars encounter sent Mars into the higher orbit where the planet moves today. After centuries of instability, beginning with the expulsion of Venus from Jupiter before 1500 B.C., the solar system finally reached a condition of relative calm as it appears today. (For a more detailed overview, see Appendix B.)

The mysterious made plain

Velikovsky's cosmological explanation of the past thus becomes a key to understanding much of that which is found in ancient documents, such as the Old and New Testaments of the Bible. Indeed, that which seemed mysterious and incomprehensible in myth, legend, folklore, ritual and even prophetic writings, is now rendered intelligible.

". . . they have made them a molten calf, and have worshipped it, and have sacrificed thereunto, and said, These be thy gods, O Israel, which have brought thee up out of the land of Egypt."

Exodus 32:8

4

Answers to Some Puzzling Questions

Velikovsky's thesis concerning ancient heavenly manifestations may explain why nearly all ancient religions commonly reverenced two images: The cow and the serpent. It may also serve to increase our understanding of scriptural references to these two objects. Although the Prophets decried idol worship, they repeatedly used imagery of both the serpent and the cow: the symbols of the ox and the snake were both used by Moses according to the Lord's instructions.

They worshipped what they saw

The worship of the cow or bull may have begun with the appearance of Venus after its expulsion from Jupiter.

According to Velikovsky, many ancient civilizations referred to Venus as once having been a comet of great and imposing appearance. For an indeterminate period of time prior to the Exodus, Venus moved on an elongated elliptical orbit that carried it across the paths of the inner planets between Jupiter and the sun. In that orbit, it must have been an imposing sight—a sign in heaven—with a tail that sometimes seemed to stretch the entire length of the sky! Apparently, from time to time, the comet appeared to take the form of a bull's head. Velikovsky wrote:

> . . . the coma [tail] of Venus changed its form with the position of the planet. When the planet Venus approaches the earth now, it is only partly illuminated, a portion of the disc being in shadow; it has phases like the moon. At this time, being closer to the earth, it is most brilliant. When Venus had a coma, the horns of the crescent must have been extended by the illuminated portions of the coma. It thus had two long appendages and looked like a bull's head. (*Worlds in Collision*, p. 175.)

Aaron was a man who had worked hand in hand with the Prophet Moses. He had been called by God to be a spokesman for that Prophet. Yet Aaron condescended to fashion a golden calf for the children of Israel that they might worship it—contrary to explicit directions from Moses.

It seems curious that the children of Israel would wish to worship a golden calf, in light of the fact that God had so severely punished the idol-worshipping Egyptians. Hadn't the Israelites been delivered just as Moses had promised? Hadn't they witnessed the power of God in their deliverance? More than anyone else, didn't they have good reason to be true to God's commandments?

Velikovsky's explanation of the unusual appearance of

the comet Venus in the days before and after the Exodus seems to account, in part, for the idolatry of the Israelites: The children of Israel forged an idol in the form of the image which they had seen in the heavens, rather than worshipping the true God of Heaven who controls all the elements. The Lord Himself, speaking through the Prophet Amos hundreds of years later, rebuked members of the House of Israel for chronic idol worship. "But ye have borne the tabernacle of your Moloch and Chiun your images, the star of your god, which ye made to yourselves." (Amos 5:26.) That statement from Amos seems to leave little doubt that the image the Israelites worshipped had some relation to a star or planet. (Planets were commonly referred to as stars because of their appearance in the heavens.)

The queen of heaven

Later, the Prophet Jeremiah attempted to eradicate the despicable practices of idol worship among the exiled Jews after Jerusalem had been destroyed by the Babylonians. But the Jews were adamant in their sacrilegious, idolatrous practices. They replied to Jeremiah:

> As for the word that thou hast spoken unto us in the name of the Lord, we will not hearken unto thee.
>
> But we will certainly do whatsoever thing goeth forth out of our own mouth, *to burn incense unto the queen of heaven,* and to pour out drink offerings unto her, as we have done, we, and our fathers, our kings, and our princes, in the cities of Judah, and in the streets of Jerusalem: for then had we plenty of victuals, and were well, and saw no evil. (Jeremiah 44:16, 17; italics added.)

The "queen of heaven" that the Jews spoke of here is the

same deity that was worshipped throughout the ancient world—the comet-planet Venus. The primary aspect of the "queen of heaven" was the sacred cow, known by various titles: The celestial cow, Athene (Venus), al-Uzza, Anat, Astarte, Ishtar, Isis or Apis. (See *Worlds in Collision,* pp. 188-190.)

There seems to be little room for doubt here. Venus became the horned god of the ancient world, represented by the ox, the bull, the calf, or the cow, depending on how a particular group of people perceived the sign displayed in the heavens. A horned god which gored the world, spilling blood onto the earth, would have been considered a male deity; but when the same god fed earth's surviving inhabitants with bread from heaven (manna) and caused the streams of the earth to flow with milk and honey—in effect, suckling the people of the earth—the horned god could then also be considered a female deity—the queen of heaven. Hence most of the gods worshipped in antiquity had both a male and a female aspect. This concept was carried so far in some cases that a few of the gods were considered to be at once both male and female—hermaphrodites.

Even the Israelites, led by a Prophet of God, succumbed to the awe-inspiring heavenly manifestation. They built an idol—a golden calf—so that they might venerate the sign in heaven. Not only that, but the above statements (together with others from Amos, Jeremiah and Acts) clearly indicate that the Israelites continued to worship idols for many centuries, contrary to the counsel of the Prophets.

Prior to the presentation of Velikovsky's theories, many scholars had sought to explain the parallel development of these ideas in many religions through a process called diffusion. They speculated that a religious concept was carried from one culture to another in the course of normal, everyday interaction between cultures. In order for the idea of cattle worship—or any other religion—to spread from one culture to another, however, a respectable amount of time must elapse. Contrary to the diffusion theory, most evidence indicates that this form of idol worship originated *almost simultaneously* in all cultures. Hence, Velikovsky's explanation more accurately fits the evidence than does the theory of diffusion.

Satan and the serpent

The other object that is almost universally included in ancient religions of the world is the serpent or snake. To Christians, the serpent is usually considered a symbol for Satan and evil, but to nearly all other ancient religions, it was a form for reverence and veneration. Worship of the serpent is found in cultures in China, India, Arabia, Egypt, Northern Europe, Mexico, and Peru, just to name a few. Where does such universal symbolism come from?

According to Velikovsky, the comet of the Exodus, at one point during its passage, appeared to exist as two separate bodies. The comet's head appeared to be a god of light, while the tail appeared to be a serpent. Electromagnetic forces within them generated tremendous electrical discharges, or arcs, that leapt between the head and tail of the comet. Other discharges sizzled within the tail itself. These sudden changes of electromagnetic potential caused the comet's tail to alter its position in space. To spectators on Earth at that time, it appeared that the snake or serpent was undulating from side to side, then changing its aspect to that of a dragon with many tails or many heads. The electric arcs themselves appeared to be giant thunderbolts or mighty arrows thrown at the serpent or snake by the god of light.

This entire celestial battle scene was enacted while great meteorite showers crashed to Earth and strata within the planet shifted, groaned and made incredible noises. It must have been an impressive audio-visual display, to say the least. So impressive was it, in fact, that the whole world remembered, and venerated the serpent.

Symbols of the snake or serpent are found everywhere in ancient literature. The serpent motif is present in virtually every culture that dates from the time of the Exodus. The Bible itself is replete with such references. For instance:

> Awake, awake, put on strength, O arm of the Lord; awake, as in the ancient days, in the generations of old. Art thou not it that hath cut Rahab [meaning "contestor"], and wounded the dragon?
>
> Art thou not it which hath dried the sea, the waters of the great deep; that hath made the depths of the sea a way for the ransomed to pass over? (Isaiah 51:9-10.)

The reference here to the Exodus is unmistakable. Rahab is a poetical name for Egypt. (*The New Smith's Bible Dictionary*, p. 317.) Naturally the word representing "contestor" would be chosen to depict Egypt because the Egyptians not only sought to restrain the Israelites, but they also attempted to oppose God Himself.

Therefore, all the elements of the great celestial battle (as described by Velikovsky) are present in the above verse from Isaiah. The "arm of the Lord" as the god of light (naturally) was seen to cut Egypt and wound the serpent or dragon. It is also singled out as the agent which moved the waters aside so that the Israelites might cross over.

It was natural that the prophets of God would employ the serpent metaphor to describe Satan. All the elements necessary for a persuasive description of that evil being were available to them in the metaphor of the serpent. Incidentally, modern folklore which depicts Satan as a beast with cloven hooves, horns on his head and a long, undulating serpentine tail is a fascinating mixture of the two more ancient images of the sacred cow and the serpent, both of which were initiated by the appearance of the comet Venus. So even today the images of the past persist!

Heaven and hell

In this celestial picture which comes from Exodus times can be found the basis of two more beliefs that Christianity inherited from the past—the concepts of heaven and hell.

Velikovsky explained that as the giant comet and its tail moved away from the Earth, the battle between the god of light and the serpent appeared to end. As the celestial apparition faded from view into the darkness, it seemed to earthbound spectators that the defeated serpent had been driven underground, somewhere beneath the Earth. They believed this because for years after the great battle was seen in the sky, Earth's inhabitants thought they could hear the serpent groaning and moving beneath them. Of course, what they heard and felt were merely aftershocks—common to all earthquakes—that continued to occur for years afterward as our planet's substrata sought to settle into place once again. Moreover, those aftershocks and the accompanying noises they generated within the Earth (variously described as tremendous moans, groans and shrieks) convinced the survivors that the serpent was still there beneath the Earth, perhaps awaiting a chance to come up from the pit, ascend into heaven, and again wreak havoc upon them.

It seems only natural that the Prophets of God would employ common metaphors of destruction to describe the punishment of the wicked. "Fire" and "brimstone" were agents of the "hell on earth," created in past destructions. To the generation that immediately followed those days, such metaphors had a very real meaning. As time went on, however, succeeding generations could not appreciate the extent or the reality of previous devastations. The events began to seem more and more abstract to them, and the true meaning of such metaphors became lost in time and cultural mythology.

Today, as part of its heritage from the past, the Christian world has received its concepts of punishment for the wicked and of Satan's abode beneath the Earth from the cultural symbols of hell and the serpent, engendered from past heavenly pageants. The Prophets used the metaphor (or myth if you will) of the serpent to describe Satan, so we have come to associate one with the other.

So also we have come to associate the metaphor or myth of "the pit," or "hell," or "the underworld" with punishment of the wicked. Dante's concept of Hades in *Paradise Lost* was merely an allegorical restatement of the popular Christian thought. Although today we know there is no "underworld" here on Earth, from the time of the Exodus down to the time of Christ and on through the Dark Ages, many people lived and died firmly believing that hell existed beneath their feet, under the Earth.

The Prophets were not pagans. They understood the correct principles of the gospel. They simply used a convenient, effective, well understood metaphor or myth to create an aversion to hell and Satan among their followers.

Likewise, we who are so far removed from the ancient events that gave rise to those myths and metaphors should not be ignorant of them and the way the Prophets employed them. Like the Prophets, we can understand the correct principles behind what appear to be metaphors in scripture.

The corollary to the concept of hell being below us is the idea of heaven being "above us." God has not revealed to man where heaven is. The popular concept of heaven probably originated with the inception of planet worship. Because the "gods" appeared only in the sky, that was thought to be their domain. This concept is still so strong that the words "heaven" and "the sky" are often used interchangeably.

The serpent in the future

The metaphor of the serpent is not only used by the Prophets when they describe past events, but also when they speak of the future:

> In that day the Lord with his sore and great and strong sword shall punish leviathan the piercing serpent, even leviathan that crooked serpent; and he shall slay the dragon that is in the sea. (Isaiah 27:1.)

John, in Revelation, resorted to using the same metaphor, and perhaps the same event, as was just quoted from Isaiah. Speaking of a time when the serpent would be bound, John wrote that he "saw a star fall from heaven unto the earth: and to him was given the key of the bottomless pit." (Revelation 9:1.) The phrase "the bottomless pit" is undoubtedly a reference to the place beneath the Earth where the serpent was thought to have fled. Also, the reference to a star receiving the key to the bottomless pit is unmistakable.

John further declared:

> And I saw an angel [star?] come down from heaven, having the key of the bottomless pit and a great chain in his hand.
>
> And he laid hold on the dragon, that old serpent, which is the Devil, and Satan, and bound him a thousand years,
>
> And cast him into the bottomless pit, and shut him up, and set a seal upon him, that he should deceive the nations no more. . . . (Revelation

20:1-3.)

The Prophets were not naive: they knew that Satan was not a serpent, but the analogy was a convenient and appropriate one. Anyone knowing the true nature of past and future destructions would understand their imagery.

The use of serpent or snake symbols in discussing Satan also may account for the description in Genesis of Eve's deception by a serpent. All of the documents from which our Old Testament was taken were written by Prophets who lived after the Exodus. It may be that they wished to impress upon people's minds that Satan, the great deceiver of mankind and the enemy to God, was as real and as menacing as that fiery serpent in the heavens which had seemed capable of destroying the Earth and all mankind with it.

Apparently the imagery was successful for nearly every Prophet since then has employed it.

The imagery of parables

Closely related to the Biblical use of imagery is the use of the parable. We know from history that the Jews employed parables as major teaching tools. The Savior himself used them. A parable—a short, fictitious story illustrating a principle—invariably has deeper meanings more important than the elementary surface story. Besides its most easily understood meaning, there are always other levels of deeper meanings.

Nephi wrote of the problem of understanding the parables of the Jews:

> . . . For behold, Isaiah spake many things which were hard for many of my people to understand; for they know not concerning the manner of prophesying among the Jews. . . .

> Yea, and my soul delighteth in the words of Isaiah, for I came out from Jerusalem, and mine eyes hath beheld the things of the Jews, and I know that the Jews do understand the things of the prophets, and there is none other people that understand the things which were spoken unto the Jews like unto them save it be that they are taught after the manner of the things of the Jews. (2 Nephi 25:1, 5.)

At the time that Nephi made these observations, the Nephites were only one or two generations removed from their ancestors in Jerusalem. Nevertheless, they could not understand the writings of the Jews! The task for us is many, many times more difficult for we are scores of generations and several diverse cultures removed from those who wrote most of the ancient scriptures. To understand their parables, metaphors and imagery, we too must understand "the manner of the things of the Jews." (2 Nephi 25:5.) Velikovsky (ironically, a Jew himself) has given us an excellent clue to the writings of the Israelites and all ancient cultures.

Understanding the prophets

The book of Isaiah provides an excellent example of how important it is to understand parables that use the imagery of the past.

Why, for example, would Isaiah refer to the King of Babylon as "Lucifer, son of the morning"? (Isaiah 14:12.) It is the only time in all the scriptures that the name Lucifer is used. The word literally means "light-bearer," and almost all biblical scholars agree on what the appellation refers to:

> Likely the planet Venus, as the morning star, which appears as a morning or an evening star, according as it is west or east of the sun, returning

> to the same position about every eighteen months. It may thus be a harbinger of daylight. (*The New Smith's Bible Dictionary*, p. 218.)

Velikovsky's affirmation that Venus caused most phenomena of the Exodus may provide a clue to Isaiah's reference to the morning star as the central character in the 14th chapter. Latter-day Saints understand that Isaiah is speaking of the fall of Satan before the world was; and while this is true, a close reading of the entire chapter reveals that the Lord is predicting, through His Prophet, the fall of the King of Babylon. Velikovsky's thesis does not alter the meaning of Isaiah as we know it, but it adds yet another level of meaning. This is very typical of Jewish parables. Providing two, or sometimes three, levels of meaning may seem unusual to us, but it was a typical practice for ancient Prophets and scholars.

In the 14th chapter, Isaiah is giving a message to the King of Babylon. Speaking in a future tense—the prophetic future tense—in verses 4 through 11 he tells what will happen to the King of Babylon as if it had already transpired. Isaiah tells of the joy of the people, and of nature—that this tyrant no longer oppresses them.

Then in the oft quoted verses pertaining to the fall of Satan from Heaven, the Lord is still speaking to the King of Babylon concerning his impending fall from power. Thus we have at least two levels of meaning, the first of which seems to be a type for the second.

Lucifer—a comet turned planet

There is yet another level of meaning, for this scripture may be understood to depict the fall of Venus from being the most awesome spectacle in the heavens: from an incandescent comet with a tremendous tail, to a seemingly tame planet.

Isaiah lived around 700 B.C., several centuries after the Exodus. Nevertheless, having studied the writings of former Prophets, he understood the true nature of the great phenomena surrounding the Exodus. He even may have seen the Exodus in vision.

Reading the 14th chapter of Isaiah in light of this knowledge provides some interesting insights.

When Isaiah said, "How art thou fallen from heaven, O Lucifer, son of the morning!" he may have been referring to the once-great comet that had soared high in the heavens but subsequently had fallen to become an ordinary morning star, a common planet "cut down to the ground" so as to be barely visible above the horizon. Furthermore, the comet "didst weaken the nations!" by means of its destructive approach to the Earth. Then Isaiah spoke of its brightness and its motion in the heavens:

> For thou hast said in thine heart, I will ascend into heaven, I will exalt my throne above the stars of God: I will sit also upon the mount of the congregation, in the sides of the north:
>
> I will ascend above the heights of the clouds; I will be like the Most High. (Isaiah 14:13, 14.)

The comet seemed to defy the order which governed the stars and the planets, for it moved independently of the rest. And in rivaling the sun in brightness, it sought to "be like the Most High."

Then Isaiah declared the fall of the comet and its condemnation, saying, "Yet thou shalt be brought down to hell, to the sides of the pit." (Isaiah 14:15.) References to the Earth as "hell" and "the pit," and to the point where horizon meets sky as "the sides of the pit" are ancient mythical references, as discussed previously, probably having their origin in past destructions. Venus—or Lucifer, the light-bearer

—now appears on the horizon, or "sides of the pit" and "they that see thee . . . narrowly look upon thee, and consider thee." (Isaiah 14:15, 16.)

Now the planet Venus can only be seen with the naked eye a few months out of the year, and then only in the morning or evening. Hence those who look at it "narrowly" see it.

Then, in what seems to be a reference to the terrible destruction visited upon the Earth by Venus, Isaiah stated that men shall inquire, "Is this the man that made the earth to tremble, that did shake kingdoms; that made the world as a wilderness, and destroyed the cities thereof; that opened not the house of his prisoners?" (Isaiah 14:16.)

And finally Isaiah may have referred to the fall of Venus from its position among the other planets—from a course high in the zenith to an orbit which leaves it barely visible on the horizon:

> All the kings of the nations, even all of them, lie in glory, every one in his own house.
>
> But thou art cast out of thy grave like an abominable branch, and as the raiment of those that are slain, thrust through with a sword, that go down to the stones of the pit; as a carcass trodden under feet. (Isaiah 14:18, 19.)

Because of a parable's capacity to convey several levels of meaning, these excerpts refer not only to the fall of Satan from his former glory, and the then impending fall of the King of Babylon, but also to the literal fall of the comet Venus.

It appears that an understanding of Velikovsky's reconstruction of history may allow us to understand more fully what the Prophets have written. And understanding the past events of the scriptures will allow us to better understand the prophecies of the last days, for the Prophets used the same metaphors and similes in speaking of the future that they used in speaking of the past. In other words, prophecy is a parable for modern man.

Astronomy, astrology and idolatry

Because several close planetary encounters have occurred in the past, we find that at various times man has worshipped the several agents of those destructions. These are the planetary gods, worshipped by pagans (as they were called by early Christians). The Prophets called such pagans "idol worshippers," because they made and reverenced images or idols of the planetary gods.

The ancients anxiously followed the movements of heavenly bodies, knowing that irregular planetary motions could portend renewed cataclysms on Earth. That fear of the planets may explain why astronomy was so important to the ancients. Their temples, pyramids, and great monoliths served as astral observatories as well as centers for religious rites and rituals. Much time and labor was spent in constructing those tremendous structures and in watching the planets march across the sky. Those who best understood astronomy rose to a high status in such communities.

This may also be the reason why the religions of the ancients were so intimately associated with astronomy. Celestial bodies were the focus of ancient religion. Fear of renewed destruction wrought by the planets/gods impelled men to render obeisance to the images that represented those fearful forces, and over time human characteristics were attributed to the celestial gods and those images. Moreover, religious rites and rituals were intimately related to past celestial events. For example, covering one's self with ashes as did the Israelites, or painting one's body red as some

American Indians did, or sacrificing humans in bloody rituals as pagans did the world over, or rituals of "passing through the fire," all entail re-enactments of elements of past destructions. In addition, the ancients exhibited intense interest in the movement of the planets; hence, the astronomers also served as the religious directors, or priesthood of their communities.

Where the Prophets held sway with their teachings, the line between worship of God and cosmological idolatry was clearly delineated. Abraham worshipped the true God, and yet he was a great astronomer, as witnessed by his writings. But where the people rejected the Prophets, the line between religion and cosmology became less distinct. Much to the distress of holy men of God, the planets themselves became objects of worship, along with the natural elements which became the destructive agents of the "celestial gods." Rather than perceiving the planets as *agents* of the Creator, the idol worshippers chose to worship the planets *themselves* as gods. This was an integral part of the mission of the Prophets: to teach and preach against the perverted worship of planetary gods (in the form of idols).

Astrology also appears to have originated from ancient cataclysms. Astrologers attempted to augur the fate of man from the heavenly positions of the planets. This logic seems reasonable considering that in past encounters man's destiny was profoundly altered by the direct intervention of the planets. In the time since the last planetary encounter, astrology has degenerated to become a parlor game, an amusing pastime, but at one time it was a form of worship. Naturally, ancient astrology was steeped in mysticism and mystery.

It seems safe to say that without the guidance provided by the Prophets, the ancients mistakenly looked for guidance to the planets rather than to God Himself, the great Architect of the Universe. In so doing, they deluded themselves and subsequently fell into futile, perverted and self-destructive practices.

"And thus the face of the whole earth became deformed, because of the tempests, and the thunderings, and the lightnings, and the quaking of the earth."

3 Nephi 8:17

5

The Book of Mormon and Catastrophism

In addition to the cataclysms described in the Old and New Testaments, the Book of Mormon contains numerous references to catastrophic phenomena. For example, the Nephite Prophets said much concerning the judgments which would attend the birth and death of the Savior. Nephi (the son of Lehi) was the first to mention what would befall his people. He quoted the Jewish Prophet Zenos as saying that three days of darkness would be the sign of the death of the Son of God.

> For thus spake the Prophet: The Lord God surely shall visit all the House of Israel at that day, some with his voice, because of their righteousness, unto their great joy and salvation, and others with the thunderings and the lightnings of his

> power by tempest, by fire, and by smoke, and vapor of darkness, and by the opening of the earth, and by mountains which shall be carried up. (1 Nephi 19:11.)

Thus, because of the similarity of the Book of Mormon catastrophe to those examined elsewhere in this book, it is particularly interesting, in light of Velikovsky's thesis, to note the signs attending the Savior's crucifixion.

The signs of catastrophe

Samuel the Lamanite was one Prophet called as a harbinger of the destructions to attend the Savior's death.

> Yea, at the time that he shall yield up the ghost there shall be thunderings and lightnings for the space of many hours, and the earth shall shake and tremble; and the rocks which are upon the face of this earth, which are both above the earth and beneath, which ye know at this time are solid, or the more part of it is one solid mass, shall be broken up;
>
> Yea, they shall be rent in twain, and shall ever after be found in seams and in cracks, and in fragments upon the face of the whole earth, yea, both above the earth and beneath.
>
> And behold, there shall be great tempests, and there shall be many mountains laid low, like unto a valley, and there shall be many places which are now called valleys which shall become mountains, whose height is great.
>
> And many highways shall be broken up, and many cities shall become desolate. . . .
>
> And behold thus hath the angel spoken unto

> me; for he said unto me that there should be thunderings and lightnings for the space of many hours.
>
> And he said unto me that while the thunder and the lightning lasted, and the tempest, that these things should be, and that darkness should cover the face of the whole earth for the space of three days. (Helaman 14:21-24, 26-27.)

Samuel also declared that "The sun shall be darkened and refuse to give his light unto you; and also the moon and the stars." (Helaman 14:20.)

Confusion of day and night

It is also interesting to note what he said concerning the sign of the birth of Christ—that it would be a day and a night and a day which would be "as if it were one day and there were no night." (Helaman 14:4.) He made it very clear that there would be, however, no disruption of the Earth's rotation.

> Therefore, there shall be one day and a night and a day, as if it were one day and there were no night; and this shall be unto you for a sign; for ye shall know of the rising of the sun and also of its setting; therefore they shall know of a surety that there shall be two days and a night; nevertheless the night shall not be darkened; and it shall be the night before he is born. (Helaman 14:4.)

It may well be that Samuel emphasized this point because he was trying to indicate a small, but significant, difference between the events to accompany the birth and death of the Savior and those of former eras (i.e., the Exodus, the long day of Joshua and the days of Isaiah), when the Earth's rotation had been altered or halted. Though the survivors of prior disruptions were confused as to the amount of time that passed while the Earth stood still, Samuel seems to have taken care to indicate to the Nephites that there would be no such confusion at that time even though the destructions would be reminiscent of past destructions.

It took them by surprise

The Book of Mormon then recounts that "it came to pass in the thirty and fourth year, in the first month, on the fourth day of the month, there arose a great storm, such an one as never had been known in all the land." (3 Nephi 8:5.)

The Nephites had never experienced the type of cataclysm which occurred at the time of the crucifixion. They therefore did not understand the metaphors and similes used by the Prophets in discussing such phenomena (2 Nephi 25:5.) As a result, they were unprepared for what befell them. As Nephi likened the scriptures to his people, perhaps we would do well to liken the experience of the Nephites to our condition, for we too have been forewarned by prophecy as the Nephites had been.

A cataclysmic replay

The Book of Mormon account tells what befell them in the day of judgment:

> And there were exceeding sharp lightnings, such as never had been known in all the land. . . .
>
> And many great and notable cities were sunk, and many were burned, and many were shaken till the buildings thereof had fallen to the earth, and the inhabitants thereof were slain, and the places were left desolate. (3 Nephi 8:7, 14.)

Although the destructions which befell the Nephites do not conform in all particulars to those described in Exodus, the similarities are many, and the pattern is nearly identical. The dissimilarities between the two events are easily explained.

As Samuel prophesied, there was no disruption of the Earth's normal rotation during this upheaval. Neither was there any precipitation of red dust, turning the waters to blood. There was sufficient ash in the air to blot out the sun for three days, but it seems that the ash, dust and smoke were mostly volcanic, not interplanetary in origin, and most likely came from volcanoes created in the tremendous shifting of the Earth's crust. The fact that Nephi pointed out that it all began with a "great storm" and then escalated to what he called "a great and terrible tempest," without first mentioning dust and ash, would seem to indicate that the dust and ash came afterwards and were solely volcanic in origin.

The plague of darkness in Egypt at the time of the Exodus is described by rabbinical sources in terms that are nearly identical to those used in Third Nephi in describing the destructions among the Nephites at the Savior's death. Although these two events were separated by approximately 1,500 years, the similarities between the two accounts is remarkable. This and other aspects of the Book of Mormon account bear a striking resemblance to Old Testament accounts of earlier destructions.

Another near collision?

Despite the dissimilarities between the events of the Exodus and the calamities associated with the crucifixion, it is possible that the phenomena experienced by the Nephites were occasioned by a type of planetary encounter. An interplanetary body without a tail or even an atmosphere would not inject dust or ash into the Earth's atmosphere, but its *gravitational attraction* would cause major disruptions of Earth's crust resulting in all the geological phenomena described in the Nephite account. Note also that the effects of this disruption were less pronounced in the Old World, perhaps indicating that the intruder passed not only quickly, but in sync with the Earth's rotation so as to directly affect only one hemisphere. Nephi's comment that the destruction in the North was worse than in the South would also tend to substantiate the idea that the invader passed in sync with Earth's rotation, thereby directly effecting only a portion of the globe.

The fact that the Earth's rotation was unaffected might have been due to the smaller size of the passing body, or its having passed at a greater distance. Of course, if it passed in sync with Earth's rotation, the tendency to alter Earth's rotation would also have been greatly minimized.

However, lest we underestimate the magnitude of the events described in the Book of Mormon, we should note that the earthquakes were not a localized phenomenon. The New Testament laconically describes the scene on the other side of the world following the crucifixion: "and there was a darkness over all the earth until the ninth hour. And the sun was darkened." (Luke 23:44-45.) Matthew added that "the earth did quake, and the rocks rent." (Matthew 27:51.) Apparently the manifestations in Jerusalem were much more subdued than those in Zarahemla, though they were still profound. The duration and magnitude of the tremors indicate more than just a large earthquake. This earthquake was apparently global in scope. A disruption of that magnitude may not easily be explained by a planet's internal forces. An external force would seem more likely as the cause of the great destructions and geologic changes described in the Book of Mormon.

Another striking similarity of the Book of Mormon account with Old Testament accounts of earlier destructions is the statement that most of the cities were destroyed and left desolate. Velikovsky noted that such encounters left entire civilizations in ruins. Ancient accounts tell of terrible bolts of lightning and thunder, earthquakes, and tempests. It appears that the Nephites were no exception to this pattern.

> . . . for behold, the whole face of the land was changed, because of the tempest and the whirlwinds and the thunderings and the lightnings, and the exceeding great quaking of the whole earth." (3 Nephi 8:12.)

Nephi made it plain

Looking elsewhere in the Book of Mormon to substantiate the theory of catastrophism, we find that Nephi (the son of Helaman) made a very interesting statement concerning phenomena associated with disruptions of the rotation of the Earth:

> For behold, the dust of the earth moveth hither and thither, to the dividing asunder, at the command of our great and everlasting God.
>
> Yea, behold at his voice do the hills and the mountains tremble and quake.
>
> And by the power of his voice they are broken up, and become smooth, yea, even like unto a valley.
>
> Yea, by the power of his voice doth the whole earth shake;
>
> Yea, by the power of his voice, do the foundations rock, even to the very center.
>
> Yea, and if he say unto the earth—Move—it is moved.
>
> Yea, if he say unto the earth—Thou shalt go back, that it lengthen out the day for many hours

> —it is done; and thus, according to his word the earth goeth back, and it appeareth unto man that the sun standeth still; yea, and behold this is so; for surely it is the earth that moveth and not the sun.
>
> And behold, also, if he say unto the waters of the great deep—Be thou dried up—it is done.
>
> Behold, if he say unto this mountain—Be thou raised up, and come over and fall upon that city, that it be buried up—behold, it shall be accursed. (Helaman 12:8-17.)

In all of scripture there is no better analysis of the geophysical and cosmological relationships seen to occur as the rotation of the Earth is disrupted. Nephi used no similes, no metaphors—just plain language. He understood that the Earth's rotation had been disrupted in the past. He even listed it with the other natural effects of disrupted rotation, i.e., rifts in the Earth's crust, the collapse of mountains, and tremendous earthquakes resulting from the wrenching of the entire lithosphere.

Where did Nephi learn of these things? Most likely he read and understood the nature of ancient events as told on the Brass Plates. In addition, he had been enlightened as to the nature of such events through revelation. Remember, he wrote the above words before the destructions enumerated in 3rd Nephi, so he was not describing the effects of those events. His description more nearly fits events that we read about in Joshua when the sun and the moon stood still. (This seems to verify the fact that Nephi got his basic facts from the Brass Plates, which contained accounts of both the Exodus and the Long Day of Joshua.)

Science says it could happen

One of the most awesome effects generated when a

body nearly collides with the Earth is the slowing and even stopping of the Earth's rotation. Velikovsky maintained that this is one of the prime elements of the collision scenario. Most scientists dismiss such an effect as pure fantasy, stating that if the Earth were ever to stop rotating suddenly, the disruption would throw people, cars and buildings off into space. But Velikovsky was adamant. He insisted that ancient records were replete with eyewitness accounts of the Earth stopping in place and even rotating backwards.

Recently several scientists (among them the noted astronomer Dr. Carl Sagan) have conceded that if the Earth's rotation somehow could be slowed gradually—over the course of an hour or so—it would not result in the fragmentation of the planet, though it would cause great disruptions. (See *The Age of Velikovsky,* p. 230.) Sagan's comments are included in an appendix to a paper dealing with Velikovsky's theories which Sagan delivered at the 1974 meeting of the American Association for the Advancement of Science (AAAS). Although Sagan is numbered among Velikovsky's greatest opponents and detractors, this concession—small though it may be—is evidence of the gradual erosion of the traditional uniformitarian position.

Knowledgeable astrophysicists will now admit that if rotation could be stopped gradually, Earth's inhabitants would not even sense the deceleration because it would be so gradual, although people would suffer the secondary effects of such a disruption—earthquakes, worldwide volcanic activity, massive heating, and so on. Of course, these are the very signs of catastrophe that are found in the ancient records and legends of cultures from around the world. The connection seems so obvious, one may express wonder and amazement that so few have seen the parallels.

Nephi said it does happen

The Book of Mormon, as with all authentic records of antiquity, contains accounts of past catastrophes. It also seems that Nephi understood that the Earth had stopped rotating during past cataclysms, and in a manner that is characteristic of the Nephite Prophets, he plainly set forth what he understood. The Book of Mormon surely substantiates the theory of catastrophism.

"And unto you it shall be given to know the signs of the times, and the signs of the coming of the Son of Man; . . ."

Doctrine & Covenants 68:11

6

Modern Prophecy—Echoes of the Past

If the idea of catastrophism—the cataclysmic instigation of cosmic and terrestrial change—is a correct principle, and if Velikovsky's cosmological explanation for the cause of such change is also correct, then we may expect these theories to conform not only to ancient scripture, but to modern revelation as well.

A great sign

While speaking at the April Conference of the Church in Nauvoo in 1843, the Prophet Joseph Smith made a statement which (in the light of Velikovsky's thesis) is most interesting.

> It is not the design of the Almighty to come upon the earth and crush it and grind it to

> powder. . . .
>
> There will be wars and rumors of wars, signs in the heavens above and on the earth beneath, the sun turned into darkness and the moon to blood, earthquakes in divers places, the seas heaving beyond their bounds; then will appear one grand sign of the Son of Man in heaven. But what will the world do? They will say it is *a planet, a comet,* &c. But the Son of Man will come *as the sign* of the coming of the Son of Man, which will be as the light of the morning cometh out of the east. (*History of the Church,* 5:337; italics added.)

Perhaps the most obvious aspect of this statement is that the Prophet used the same terms to describe a future agent of destruction that Velikovsky used to describe the agent of past catastrophes—a body which looked and behaved like a comet, but was large enough to also be called a planet. The Prophet's statement indicates some confusion among those who are to see this "grand sign," as to whether it is a comet or a planet. According to Velikovsky, the intruder of the Exodus was planet-sized, even though it looked and behaved like a comet—leading to some confusion of terms in ancient records. So both the ancient records and modern prophetic statements seem to have the same problem with terminology.

Important terminology

This confusion, however, is not a drawback, because it serves to emphasize the striking similarity between past and future events. What is more, this apparent confusion of terms serves, in reality, as an excellent description of the intruding body, since it will have the characteristics of both a comet (because of its luminosity, its tail and its elliptical orbit), and a planet (because of its size). The use of both terms in the Prophet's statement gives a clear indication of the nature of the "grand sign," and lends credibility to Dr. Velikovsky's thesis.

In addition, the planet/comet (the cause) is directly associated by the Prophet with the signs of destruction (the effects). The signs enumerated by the Prophet—the sun turning dark, the moon turning red, earthquakes, oceans sweeping over land masses, etc.—are identical to the phenomena associated with the events of the Exodus which Velikovsky attributed to just such a comet-like planet.

According to Joseph Smith, when the planet or comet appears, the world will refuse to recognize it as a sign. Though earth's viewers will not regard it for what it truly is, yet, how could a comet/planet elicit any special notice or comment unless it were doing the unexpected—the unusual? Since Joseph Smith referred to it as a "grand" sign, it may be safe to suppose that it will be somewhat spectacular. And what could be more spectacular than another planet or comet passing within a very short distance of the Earth?

This and similar statements attributed to the Prophet Joseph Smith seem to indicate that he understood the nature of impending catastrophes in the same light as those Velikovsky proposed for past catastrophes. Yet, these two men were separated by more than an entire century. Velikovsky reached his conclusion through researching the writings of ancient cultures, the Prophet obtained his knowledge through revelation—yet the conclusions of each seem to substantiate those of the other.

Another interesting aspect of the statement made by the Prophet is the initial reference to the fact that "it is not the design of the Almighty to come upon the earth and crush it and grind it to powder." The Lord has declared that were Elijah not to come first, "the whole earth would be utterly wasted at his [the Lord's] coming." (Doctrine & Covenants

2:3.) We may take heart that Elijah has come and that the Lord does not wish to destroy the Earth. Nevertheless, such statements are indicative of the extent of the upheavals which will precede the Second Coming.

Hints in modern revelation

Because of the lack of understanding that exists today with regard to the theory of catastrophe, we miss the meaning of many off-hand statements made in scripture.

For example, during the time the Prophet Joseph Smith was interned in Liberty Jail, the Lord informed him that "God hath set his hand and seal to change the times and seasons." (Doctrine & Covenants 121:12.) How, we might ask, could such a change be brought about? Could the Earth's rotation or declination be altered? Perhaps the length of its orbit about the sun could be changed. Or perhaps it could be moved from its present orbit altogether!

In our day, season follows season, and the hours of the day (the times) are the same year after year. It would seem that it has always been so, for we have known nothing else for nearly two millennia. Yet, if we had lived anciently, in "the day of the Lord's wrath," we might have experienced just such a change. According to Velikovsky, many ancient sources recorded that the seasons have been changed. "The winter is come as summer, the months are reversed and the hours are disordered." (*Papyrus Anastasi IV*, as quoted in *Worlds in Collision*, p. 135.) Another source declared that "the breath of heaven is out of harmony. . . . The four seasons do not observe their proper times." (*Texts of Taoism*, I, 301, as quoted in *Worlds in Collision*, p. 135.) And yet another states that

> the thickened air concealed the heaven from view, and the stars were confused with a disorderly huddle of fire and moisture and violent fluxions of

> winds. The sun was not fixed to an unwandering and certain course, so as to distinguish orient and occident, nor did he bring back the seasons in order. (Plutarch, *Morals.*)

Finally, an ancient text from Mexico states that "the order of the seasons was altered at that epoch." (Brasseur, *Sources de l'histoire primitive du Mexique,* pp. 28-29.)

Velikovsky explained that

> with the fall of the [Egyptian] Middle Kingdom and the Exodus, one of the great world ages came to its end. The four quarters of the world were displaced, and neither the orbit nor the poles nor, probably, the direction of rotation remained the same. The calendar had to be adjusted anew. The astronomical values of the year and the day could not be the same before and after an upheaval in which, as the quoted Papyrus Anastasi IV says, the months were reversed and 'the hours disordered.' (*Worlds in Collision,* p. 135.)

Thus, it seems natural that the Lord should inform Joseph Smith that one of the changes incident to the Second Coming is to be an alteration in "the times and seasons." We can expect to see the current order of things vastly changed by the time the Savior appears again to the world. The beginning of the Millennium will undoubtedly be remembered as the end of one world age and the beginning of another!

The end of peace

Furthermore, in Section One of the Doctrine & Covenants, the Lord declares that "the hour is not yet, but is nigh at hand when peace shall be taken from the earth." (Doctrine & Covenants 1:35.) This revelation was received in 1831. Of course, the Civil War was then yet future, as were two world wars and innumerable others, but incessant wars have plagued the Earth for centuries. It almost seems unreasonable for the Lord, in 1831, to speak of taking peace from the Earth when there had been no peace to speak of for centuries. War would be nothing new—even the terrible destruction of modern warfare. What else could he have been speaking of?

It is very likely that the Lord, speaking through Joseph Smith, was referring to the cataclysms associated with the Second Coming. This planet and its inhabitants have enjoyed a prolonged period of peaceful continuance wherein nothing has disrupted the day-in-day-out rotation of the planet for centuries. The prophesied destructions would bring an end to the Earth's quiet existence. If the times and seasons are to be changed, then the accompanying disruptions would certainly take peace from the Earth.

These two statements made by the Lord to Joseph Smith are not simply poetic phrases. They are indicative of major disruptions to be caused by the "great sign." Such seemingly simple, poetic phrases can be found throughout scripture, and may easily be overlooked or misunderstood.

Modern revelation

Of course, the Doctrine & Covenants contains numerous warnings and descriptions of latter-day destructions, but given Velikovsky's thesis, the phenomena spoken of in those passages seem more comprehensible. For example:

> For not many days hence and the earth shall tremble and reel to and fro as a drunken man; and the sun shall hide his face, and shall refuse to give light; and the moon shall be bathed in blood; and the stars shall become exceedingly angry, and shall cast themselves down as a fig that falleth from off a

> fig tree.
>
> And after your testimony cometh wrath and indignation upon the people.
>
> For after your testimony cometh the testimony of earthquakes, that shall cause groanings in the midst of her, and men shall fall upon the ground and shall not be able to stand.
>
> And also cometh the testimony of the voice of thunderings, and the voice of lightnings, and the voice of tempests, and the voice of the waves of the sea heaving themselves beyond their bounds.
>
> And all things shall be in commotion; and surely, men's hearts shall fail them; for fear shall come upon all people. . . .
>
> And immediately there shall appear a great sign in heaven, and all people shall see it together. (Doctrine & Covenants 88:87-91, 90.)

The clarity and accuracy of this description is remarkable. Perhaps nowhere else in scripture is there a better cataloging of the phenomena occasioned by a planetary close encounter. This passage even refers to the "great sign" which Joseph Smith said would be called "a comet, a planet."

Even the admonition of the Lord for the Saints to watch for his advent in heaven carries a similar reminder of what will occur just before he appears again:

> Wherefore, be not deceived, but continue in steadfastness, looking forth for the heavens to be shaken, and the earth to tremble and to reel to and fro as a drunken man, and for the valleys to be exalted, and for the mountains to be made low, and for the rough places to become smooth—and

all this when the angel shall sound his trumpet. (Doctrine & Covenants 49:23.)

Restored knowledge—an affirmation

Notwithstanding their clarity and simplicity, the Doctrine & Covenants and other modern revelation contain the same metaphors used by the Prophets of old. Attempting to understand revelation without a clear concept of the meaning of prophetic metaphors may be to attempt the near impossible. As with all such parables, there is more meaning than that which is apparent on the surface. Therefore, the more understanding we have, the more meaning we will find in the parables of prophecy.

(See Appendix C for a dream or vision of the last days given to Patriarch Charles Evans prior to 1894. It is an excellent example of how the principles propounded by Velikovsky provide the key to understanding prophecy.)

"And I will shew wonders in the heavens and in the earth, blood, and fire, and pillars of smoke. The sun shall be turned into darkness, and the moon into blood, before the great and terrible day of the Lord come."

Joel 2:30, 31

7

The Book of Revelation— A New Perspective

Through the Book of Mormon we have a special witness that the writings of John (as recorded in the Book of Revelation) are true (at least insofar as they have been correctly translated). This witness stems from the fact that John's vision was part of the same revelation given to Nephi—that part which Nephi was commanded not to write. It follows that we ought to become familiar with John's record. But Revelation seems to be an enigma whose symbolism and obscure references make it difficult to understand.

John used the words of Moses

A quick review of the Book of Revelation reveals that

much of the language which John used is similar to that used in discussing the phenomena of the Exodus. In fact, it almost sounds as though John employed the plagues of the Exodus right out of the Old Testament. (See Chapter 2 for a side by side comparison of sample texts from Exodus and Revelation.) Nevertheless, even though John employed ancient usages, there is no doubt that he was writing about the future: "Write the things which thou hast seen, and the things which are, and the things which shall be hereafter." (Revelation 1:19.)

Since it is true that the Prophets employed the same terminology whether discussing past or future destructions, then an examination of the Book of Revelation in light of the Exodus account (and the light which Velikovsky's thesis seems to throw upon that account) may help to decipher what otherwise is, for most readers, an unintelligible prophecy.

The future is a replay of the past

As pointed out before, the destructions and plagues enumerated in Revelation are grouped in much the same order as those discussed in Exodus and elsewhere in the Old Testament. It is this grouping of plagues which (throughout the scriptures) indicates a planetary encounter as the cause of the phenomena mentioned. And the grouping of these plagues is important. A planetary encounter such as that which Velikovsky proposed for the Exodus would have specific effects, and a valid eyewitness account should, of necessity, mention those effects.

Revelation is extremely accurate in its grouping of plagues. John has written so that anyone who understood the nature of past upheavals might comprehend the metaphors which he employed. John saw the same things occurring in the future which had happened in the past, and he was very careful to describe future events in the same language which earlier Prophets had used in describing events of past ages.

The only thing for which John could not compensate was the ignorance of those living in the last days who would someday read his writings—those, as Peter said, who would be ignorant of the fact that the heavens and the Earth had been dramatically changed in times past, and would be so changed again. (II Peter 3:3-7.)

The beginning of sorrows

The first mention found in Revelation of the calamities that are to befall the Earth is in Chapter 6:

> And I beheld when he had opened the sixth seal, and, lo, there was a great earthquake; and the sun became black as sackcloth of hair, and the moon became as blood;
>
> And the stars of heaven fell unto the earth, even as a fig tree casteth her untimely figs, when she is shaken of a mighty wind.
>
> And the heaven departed as a scroll when it is rolled together; and every mountain and island were moved out of their places. (Revelation 6:12-14.)

Earthquakes

The first feature of this event is that of the earthquake. Although earthquakes occur throughout the world on a fairly regular basis, it appears that since "every mountain and island" will be moved, this is to be no localized earthquake, but one global in scope.

What could cause the entire Earth to quake? None of the forces we see in operation around us today seem great enough to occasion such an earthquake. Yet, if Velikovsky's thesis is correct, such a disturbance is not only possible but also a matter of historical reality.

Falling stars

Another common reference in prophetic narratives of events of the last days—a reference also found in Revelation—is that made to stars falling. Velikovsky explained that in the past when the Earth encountered another body in space, the poles of the Earth shifted position, i.e., the Earth tipped due to the forces unleashed by the proximity of the two bodies. To an inhabitant of the Earth, it would appear that the star field overhead had fallen below the horizon and disappeared. This might also account for the use of the term "new heaven." The old star field in the sky would have been replaced by a new star field, rising as the old one fell. Such a sidereal change might also account for the statement: "the heaven departed as a scroll when it is rolled together." In fact, in light of such an explanation, the simile is most vivid, for it leaves little doubt as to what John saw in vision and what will be seen in the last days.

Darkness

Due to the earthquakes and the tidal effect created by the body's gravitational pull as it approached the Earth, the Earth's molten core would be affected so as to dramatically increase volcanic activity on the Earth's surface. The atmosphere would be darkened with ash thrown from many new-born volcanoes, causing the sun to become "black as sackcloth of hair." (Revelation 6:12.)

Incandescent moon

The moon would be a scene of great heating because the electromagnetic fields of the Earth and the planet-comet would also interact with the moon's electromagnetic field. Such interaction would generate heat within all three orbs by electromagnetic induction. Lacking both atmosphere and oceans to dissipate the heat (surrounded by the vacuum in space acting as an insulator rather than a conductor), the moon would simply become incandescent. Hence, the Prophet declared that "the moon became as blood." Notice that he did not say that it *became* blood, but he used a simile, essentially saying that it turned red. (Revelation 6:12.)

As discussed earlier, these signs are all well grouped together and portray the natural effects of a planetary encounter. John seems to have listed these signs carefully so that the reader will not misunderstand the nature of the event he has seen in vision.

Inhabitants hide

It should also be remarked concerning the cataclysm which John saw, that the Earth's inhabitants hid themselves out of fear for what they were experiencing.

> And the kings of the earth, and the great men, and the rich men, and the chief captains, and the mighty men, and every bondman, and every free man, [virtually everyone] hid themselves in the dens and in the rocks of the mountains. (Revelation 6:15.)

Some have said that this urge to hide is a natural, instinctive impulse when one is caught in an earthquake. But it is a well known fact that people commonly do not hide in an earthquake. They instinctively run into the open (although in a city, the street may be the least safe place to be because of falling debris). John stated that during this catastrophe the people hid themselves. The desire to hide under something would be engendered by fear of destruction from above. From what, then, might they have been hiding if not from an earthquake? Perhaps they sought refuge from a fearful object overhead or impending destruction coming from above. It is not unlikely that people sought refuge in caves and overhanging escarpments during past celestial catastrophes. That would account for rich archaeological caches of primitive tools excavated from the floors of inhabitable caves. According to John, such places will again be sought for safety.

Moreover, John implies that the people would have time to flee, a condition precluded by an earthquake. There is no warning of an impending earthquake, but an approaching body from space could be seen by everyone. There would be some time to hide.

The implication is clear. If Velikovsky's explanation is correct, then John seems be saying that the Earth will be visited by another body from space which will cause the Earth's crust to crack and buckle, the poles to shift, the atmosphere to blacken and the moon to heat to incandescence.

Trumpets

Then, in the eighth Chapter of Revelation, we are introduced for the first time to the seven angels with trumpets who will herald each of the upcoming plagues. But before examining the plagues and destructions meted out by the angels, it may be informative to look at the activity of the angels themselves.

Among the most prominent features of the signs disclosed in Revelation is the sounding of trumpets by various angels, each heralding a new plague or cataclysm to befall the inhabitants of Earth. This has always seemed largely symbolic to Christians, and because most of John's references seemed obscure, many readers have considered the entire book to be symbolic. Yet if we accept the account of the Exodus as correct, it appears that a trumpet did sound during the lawgiving at Mt. Sinai.

> And it came to pass on the third day in the morning, that there were thunders and lightnings, and a thick cloud upon the mount, and the voice of the trumpet exceeding loud; so that all the people that were in the camp trembled. . . .
>
> And when the voice of the trumpet sounded long, and waxed louder and louder, Moses spake and God answered him by a voice. (Exodus 19: 16, 19.)

Not only the Old Testament, but modern revelation as well, declares that a trump will again sound. "For a trumpet shall sound both long and loud, even as upon Mount Sinai, and all the earth shall quake." (Doctrine & Covenants 29:13.) Orson Pratt wrote that the sounding of trumpets is to be understood literally. "And the time will come, when the seven angels having the seven last trumps will sound their trumps literally, and the sound thereof will be heard among the nations, just preparatory to the coming of the Son of Man." (*Journal of Discourses,* 18:227.)

Midrashic literature tells of seven different pitches of trumpet sounds heard at Mt. Sinai. (Notice the correlation of the seven pitches with the seven angels sounding trumpets in Revelation.)

Music of the spheres

Velikovsky suggested that the sound of a trumpet, heard at Mt. Sinai, was a consequence of the comet approaching Earth, and he maintained that the sound was heard worldwide.

As two charged bodies (such as planets) approach one another in space, apparently a pure tone is generated in the atmosphere, likely by electromagnetic oscillation. As the distance between the two bodies increases or decreases, the intensity and the pitch of the sound changes—much like the operation of a Theremin. (A Theremin is an electronic musical instrument, a forerunner of modern music synthesizers, which may be used to create eerie electronic music.)

Simply another sign

Because the sound heard in connection with the law-giving at Mt. Sinai was apparently generated by the two orbs, it was heard around the world. Naturally, these trumpet-like sounds would have also accompanied the other disasters meted out by the comet. Thus, if the plagues spoken of by John are to be understood as having been caused by the approach of a planet-like comet, then the sound of trumpets mentioned in connection with the Exodus could very appropriately serve to explain the trumpets mentioned in Revelation. In other words, John was simply listing another of the physical manifestations of the near collision of the Earth with another body in space. Therefore, in a very literal manner, the sound of trumpets is to accompany the prophesied cataclysms prior to the Second Coming of the Savior.

> Wherefore, be not deceived, but continue in steadfastness, looking forth for the heavens to be shaken, and the earth to tremble and to reel to and fro as a drunken man, and for the valleys to be exalted, and for the mountains to be made low, and for the rough places to become smooth—and all this when the angel shall sound his trumpet. (Doctrine & Covenants 49:23.)

Before the angels sound their trumpets, another angel is to fill his censer with fire from the altar in heaven and cast it to the Earth. This is to be attended by "voices, and thunderings, and lightnings, and an earthquake." (Revelation 8:5.)

Velikovsky maintained that during the law-giving at Mt.

Sinai, the Earth groaned and the skies thundered. We have seen that the records of cultures around the world attest to the fact that all the world heard these sounds, not just the Israelites alone.

The destructions begin

In Chapter 8 of Revelation, the first of the seven angels sounds his trumpet, and three destructions follow: hail, fire, and blood (plagues identical to those of the Exodus): "The first angel sounded, and there followed hail and fire mingled with blood, and they were cast upon the earth: and the third part of trees was burnt up, and all green grass was burnt up." (Revelation 8:7.)

John grouped these three together saying that they were "mingled," indicating a more or less simultaneous occurrence. Again, using the account of the Exodus as a means of understanding prophetic metaphors coupled with Velikovsky's thesis, the three plagues will be meteorites, burning naphtha, and limonite or some form of ferric oxide. In succeeding verses, John indicates that the red dust will make water unpotable. It will also kill marine life and turn the water bitter—the same effects ascribed to the bloodied water of the Exodus.

Moving ahead to verse 12, we see that John also spoke of the darkening of the Earth's atmosphere:

> And the fourth angel sounded, and the third part of the sun was smitten, and the third part of the moon, and the third part of the stars; so as the third part of them was darkened, and the day shone not for a third part of it, and the night likewise. (Revelation 8:12.)

This darkening seems to be caused by the dust and debris of the comet entering Earth's atmosphere, as well as by ash thrown into the atmosphere by many new volcanoes. John's reference to a shortening of the Earth's day-night cycle may be an indication of disrupted rotation, as well. In fact, the reference "and the day shone not for a third part of it" seems to be a direct reference to a change in Earth's sidereal motion.

John points out the cause

It seems evident that John was speaking of a close encounter of a most disastrous kind. Nevertheless, the most stunning aspect of his narrative is those verses wherein he wrote of the cause of those disasters:

> And the second angel sounded, and as it were *a great mountain burning with fire was cast into the sea:* and the third part of the sea became blood. . . . (Revelation 8:8; italics added.)
>
> And the third angel sounded, and there fell *a great star from heaven,* burning as it were a lamp. . . . (Revelation 8:10; italics added.)
>
> And the fifth angel sounded, and I saw *a star fall from heaven unto the earth. . . .* (Revelation 9:1; italics added.)

The clarity of John's statements is not clouded by his use of the term "star," for the point is that the Earth will come in close contact with celestial bodies—whether we call them planets, comets, or stars. The last verse is even emphatic. There appears to be little doubt about what John saw and what he wished to communicate to those who would read Revelation.

As if that weren't enough, John went on to give even more compelling evidence for this:

> And the fifth angel sounded, and I saw a star fall from heaven unto the earth: and to him was

> given the key of the bottomless pit. And he opened the bottomless pit: and there arose a smoke out of the pit, as the smoke of a great furnace; and the sun and the air were darkened by reason of the smoke of the pit. (Revelation 9:1, 2.)

As examined earlier, the metaphor of the bottomless pit comes from a time when the people of the Earth saw a battle in the sky, wherein they imagined they saw the dragon or serpent defeated and driven underground by the god of light. (The bottomless pit was believed to be somewhere beneath the Earth, where the serpent hid.) In these verses, John pointed directly to the star (comet/planet) that would fall from heaven as the agent which is to cause the smoke that fills the air. As we learned earlier, the blackened atmosphere is one of the earliest signs of impending disaster. Again, this fits the cause and effect relationship outlined by Velikovsky. John is simply saying that the same sort of thing will happen in this future encounter. (Note how effectively John used the metaphor of the battle in the sky. Evidently, what he saw in vision appeared to match the mythical account, so he employed that useful metaphor.)

Elder Parley P. Pratt also interpreted the matter of stars falling from heaven to the Earth in the same manner. He affirmed that these "stars" would menace the Earth, causing the prophesied cataclysms of the last days. (*Millennial Star*, 1:258. See also Chapter 9, "When They Return.")

An impressive picture is now beginning to appear. What was once a baffling prophecy now seems to be an extremely accurate account of future destruction. It appears that in using imagery, John was not deliberately trying to mislead or mystify those who might read Revelation. He was actually using metaphors that were well known and well understood anciently, in an attempt to present a forceful, dynamic picture of what he had seen in vision. Therefore, the confusion experienced upon reading Revelation stems not from any attempt by John to mystify us, but from our own ignorance as to the nature of past cataclysmic events and the myths and metaphors they spawned.

Locusts? or something else?

John then noted that locusts would come out of the smoke. The fact that the locusts are accompanied by smoke (verse 3), and that they do not eat the plants of the Earth (verse 4), should indicate that these are not truly insects. John is still speaking metaphorically. (Nevertheless, he is speaking of a literal phenomenon.)

The parallel between this Chapter of Revelation and the Book of Joel is no mere coincidence. The "locusts," we learn from Joel and John, have the appearance of horses, and then we are told that the heads of the horses were as the heads of lions. A sound like chariots and horses running to battle is heard. Both John and Joel seem to be describing the same future event, or at least they are describing the same type of phenomenon. Isaiah also describes these same creatures as "the terrible ones." (Isaiah 29:5.) What are they?

Perhaps other ancient records and past events will answer that question. We can read an almost identical account in the Vedic Hymns, the sacred records of the Hindus, where these mysterious creatures are called Maruts. In Greek mythology, Homer described them as terrible creatures, the steeds of Ares (Mars).

Velikovsky claimed that these ancient records, and many others like them, were describing a time during past inter-planetary encounters when the moons of Mars were seen to accompany the planet as it passed the Earth. The conditions that existed in space at that time caused the moons of Mars to have comet-like tails. Their appearance was such that they looked like steeds, pulling the passing planet Mars as a chariot. That would serve to explain John's reference to the future "locusts" as having the appearance of horses. John's description of the horses as having "the heads of lions" may have been because their appearance was similar to that of lions' manes.

At the same time in ancient days, multitudes of small, short-period comets were created by asteroids and other space debris captured by the wandering, errant planets. For a time, these packs of asteroids and debris carried some atmosphere with them so that they appeared to be comets. Their orbit intersected with that of the Earth from time to time after the original encounter, and caused much damage and noise—keeping mankind in a nearly constant state of fear. Therefore, the "locusts" John reported were very similar to past phenomena described by Isaiah, Joel and the Vedic Hymns.

John's reference to these "locusts" as having "hair of women" (verse 8) should confirm their identity as comets! The English word "comet" comes from the Latin "coma," which means hair, an apt description of the long, flowing tail of a comet. In addition, John declares that from these creatures will come "fire and smoke and brimstone" (verse 18), familiar elements in past destructions, and further confirmation of the fact that these creatures are not animals or insects at all, but physical extraterrestrial manifestations.

In verse 11, John indicated precisely what the origin of these things was:

> And they had a king over them which is the angel of the bottomless pit, whose name in the Hebrew tongue is Abaddon, but in the Greek tongue hath his name Apollyon. (Revelation 9:11.)

Apollyon means "destroyer." *Abaddon* means "destruction." (*The New Smith's Bible Dictionary*, pp. 25, 1.) Again, John turns to the ancient legends of the serpent that sought to destroy the world to clarify the meaning of what he is describing. The "angel of the bottomless pit" was the serpent—part of the comet/planet. It appears that John wished to emphasize how much the future destructions would resemble those that occurred anciently. Apparently these things will again be seen, a profound statement for those of us living in the day and time when the events of this prophecy will unfold.

Woe, woe, woe

John also made this interesting observation in the next verse: "One woe is past; and, behold, there come two woes more hereafter." (Revelation 9:12.) Apparently, he applied the term "woe" to the destructive manifestations seen to accompany a planetary encounter. The Savior used the term "sorrows." (Matthew 24:6.) Both seem to be appropriate, all-inclusive terms, because the manifestations which accompany such catastrophes are truly woeful, bringing much sorrow and suffering upon mankind.

The destruction having abated for the moment, John speaks of spiritual things in Chapter 10. Notice, however, that only the first woe had passed. (He made it plain that two more woes were to follow.) In Chapter 11 John tells of the second woe: the two witnesses who will testify to the Jews in Jerusalem in the last days, of the destructions that are to accompany their work—including the familiar signs of blood, plagues and a great earthquake. John then wrote that "the second woe is past; and, behold the third woe cometh quickly." (Revelation 11:14.)

The second woe, like the first, was accompanied by signs which we have come to understand as the phenomena associated with a collision or near collision.

The third woe comes in Chapter 16, and is undoubtedly the worst of all. After describing the events of the third woe, John wrote, "And the seventh angel poured out his vial into the air; and there came a great voice out of the temple of heaven, from the throne, saying, It is done." (Revelation 16:17.) John left no doubt that the "woes" had ended and that the Earth would shortly enter into its rest.

An important conclusion

From John's earlier reference to three bodies colliding or nearly colliding with the Earth, and from his reference to three separate woes, it would appear that encounters with the Earth will occur at least three times prior to the Second Coming. Each time, intruding bodies from space will be sent by the great Architect of the Universe to destroy the wicked and corrupt; to dismantle all existing societies, organizations, and governments; and to prepare the Earth for the glory of the Millennium.

Whether or not the number of encounters is correct is of little significance, except perhaps as an indication that there will be more than one. Past encounters have apparently always followed this important pattern: there was never just one, single intersection of the planets. Velikovsky indicated that Venus passed perilously close to Earth approximately every 50 years after the Exodus encounter. Later, Mars encountered Earth on at least three occasions, separated by intervals of approximately 15-20 years. After the first encounter, the probability of subsequent encounters increases, according to the laws of orbital mechanics. That is, once the orbits of the planets are altered in the first encounter, they tend to remain in their new orbits, which will eventually return them to the same point in space where the initial encounter occurred. Hence, any collision scenario will usually contain repeated encounters. John's account in Revelation follows that pattern.

More details

Returning to the second and third "woes," there are some other items worth noting.

Some very special powers are to be given to the two "witnesses" who will be called upon to testify to the Jews in Jerusalem (Revelation Chapter 11). They will be able to devour their enemies with fire, stop the rain, turn water to blood and "smite the earth with all plagues, as often as they will." (Revelation 11:8.) Remember that in the account of the Exodus, miracles performed at that time were attributed by the people to the power of Moses, when in fact they were God's work. Moses was merely God's servant, his mouthpiece on Earth, announcing the plagues and warning mankind as God bade him do so. So it will very likely be with regard to the two witnesses mentioned in Revelation. They will probably not perform these miracles themselves; rather, they will be acting as agents of God.

In any event, the destructions attending the ministry of those two witnesses point to the signs of an interplanetary near collision. Notice that in verse 7, the beast—also known as the dragon or serpent—from the bottomless pit is the instrument of the witnesses' demise. Evidently they will be killed by the same agent that will bring death to others:

> And the same hour was there a great earthquake, and the tenth part of the city fell, and in the earthquake were slain of men seven thousand: and the remnant were affrighted, and gave glory to the God of heaven.
>
> . . . and there were lightnings, and voices, and thunderings, and an earthquake, and great hail. (Revelation 11:13, 19.)

The second woe is passed. Again we have seen familiar signs as John explained what he saw.

The third is the worst

The last of the three woes begins in Chapter 16. We find that the seven angels will relinquish their trumpets for "vials of the wrath of God" and proceed to pour them out, one at a time, upon the inhabitants of the Earth: "And the first went, and poured out his vial upon the earth; and there fell a noisome and grievous sore upon the men which had the mark of the beast, and upon them which worshipped his image." (Revelation 16:2.)

Remember that the Egyptians at the time of the Exodus experienced a similar plague from the ash or dust which fell from the sky and "chaffed" their skin, causing boils and blains. Even though the account of the Exodus seems to imply that this happened only to the Egyptians, it likely happened to everyone who ventured out into the descending dust. So, too, anyone who comes in contact with the "ash" will very probably become ill from it. John again used the metaphor of the "beast" to explain that those who have the "mark of the beast" (who have been contaminated by the red dust or ash) will be afflicted by this plague.

Also, it appears that people will have begun to worship the beast or dragon which has appeared in heaven, just as they did at the time of the Exodus. Ironically, the same heavenly display of the serpent or dragon that instigated widespread idolatry in ancient times will again incite the Earth's inhabitants to similar false worship.

> And I beheld another beast coming up out of the earth; and he had two horns like a lamb [ram, or ox?], and he spake as a dragon.

> And he exerciseth all the power of the first [ancient?] beast before him, and causeth the earth and them which dwell therein to worship the first beast, whose deadly wound was healed.
>
> And he doeth great wonders, so that he maketh fire come down from heaven on the earth in the sight of men,
>
> And deceiveth them that dwell on the earth by the means of those miracles which he had power to do in the sight of the beast; saying to them that dwell on the earth, that they should make an image to the beast. . . . (Revelation 13:11-14.)

The fourth angel is to cause men to be burned. Two phenomena in these destructions might cause this. The first, as we have seen, is flaming naphtha falling in large, blazing sheets or strings, and burning on ground or water, catching fire again and again. The second is the exposure to the sun of a large portion of the Earth's surface for a prolonged period as rotation slows and then stops. This prolonged day could easily cause uncomfortably high temperatures on the surface of our planet. Both these phenomena may occur in an encounter with a planetary body, so it seems likely that John was speaking of both.

The next angel is to cause darkness. Taking into account the previous sign, one complements the other. If there is a prolonged day on one side of the Earth, there must be a prolonged night on the other side. Furthermore, we have seen how prodigious amounts of dust and debris entering the atmosphere cause the sky to darken.

The sixth angel is to cause a great river to dry up. This would be a natural consequence of several things: geologic activity, rotational disruption and axis displacement. Velikovsky pointed out that such manifestations as rivers or springs drying up, flowing upstream, dramatically altering their courses, or bursting out where none existed before, are all well documented in ancient, historical accounts of cataclysms.

The seventh angel is to cause some other signs, which have been seen before:

> And there were voices, and thunders, and lightnings; and there was a great earthquake, such as was not since men were upon the earth, so mighty an earthquake, and so great.
>
> And the great city was divided into three parts, and the cities of the nations fell. . . .
>
> And every island fled away, and the mountains were not found. And there fell upon men *a great hail out of heaven,* every stone about the weight of a talent: and men blasphemed God because of the plague of the hail; for the plague thereof was exceeding great. (Revelation 16:18-21; italics added.)

Notice the size of the "hail." "The Jewish talent is estimated to be from 75-93 pounds." (*The New Smith's Bible Dictionary,* p. 416.) The weight of such "hail" is more indicative of meteorites falling than hail as we know it. It is also indicative of the terrible damage such missiles would do.

Perhaps this is the earthquake that will return the land masses (continents) of the Earth to their former places. (Doctrine & Covenants 133:24.)

As an indication of the extent of this last great earthquake, John stated in verse 20 that "every island fled away, and the mountains were not found." Islands and mountains are actually the same structures, geologically speaking—islands are nothing more than the tops of submerged mountains. During a great earthquake caused by an astral body, it seems appropriate to expect to see new mountains born as the Earth's crust ruptures and buckles, but John says that mountains and islands will disappear. Such a change of topography, of necessity, would involve the entire Earth and would call for forces several orders of magnitude greater than any before experienced.

Last of all, John wrote that the hail which will accompany the last "woe" will be the most frightening of all, calling it "exceeding great." (The Old Testament states that the sound of the meteorites falling upon Egypt was more fearsome than the damage which they actually did.)

John wrote plainly

It appears that John has painted an accurate, if unpleasant, picture of the destructions of the last days. In fact, Joseph Smith said that "the book of Revelation is one of the plainest books God ever caused to be written." (*History of the Church,* 5:342.) From what we have considered, John's account does seem much more understandable.

Of course, the only way to fully understand the book is through the Spirit, as Nephi taught. For as he wrote, the words of Isaiah (and certainly John's words) "are plain unto all those that are filled with the spirit of prophecy." (2 Nephi 25:4.) Nevertheless, the key which Velikovsky's explanation provides appears to be of great value to the student of the scriptures.

"Behold, he cometh with clouds; and every eye shall see him, and they also which pierced him; and all kindreds of the earth shall wail because of him. Even so, Amen."
Revelation 1:7

8

The Siege of Jerusalem

As a people, Christians seem fascinated with the re-establishment of the nation of Israel after the close of World War II. Latter-day Saints, in particular, have shown an intense interest in the affairs of the Jews from the very earliest days of the restoration of the gospel. Perhaps that is because prophecy dealing with the Savior's return also tells of the return of the Jews to Jerusalem in the last days, prior to the Second Coming.

Profoundly significant was the mission undertaken by Orson Hyde, commissioned by Joseph Smith, to travel to Palestine, where (in 1841 on the Mount of Olives) he dedicated the land to the return of the House of Israel. Shortly thereafter, the spirit of gathering fell upon the Jews who began returning to their homeland in ever increasing numbers until in 1948 the United Nations created the State of Israel out of the Arab nation of Palestine. Thus, the establishment of modern Israel in the Middle East was in fulfillment of prophecy and an indication that the Savior's advent is imminent.

Many, many prophecies exist that tell about the future of Jerusalem and the things that are to befall that city in the last days. Those prophecies are particularly fascinating in light of Velikovsky's thesis.

Jerusalem will stand alone

The Old Testament Prophets spoke of a time when "all nations" will lay siege against Jerusalem. Zechariah, for example, gave a fascinating and detailed account of what will happen during a great battle for that holy city in the last days, proclaiming: "Behold, I will make Jerusalem a cup of trembling unto all the people round about, when they shall be in the siege both against Judah and against Jerusalem." (Zechariah 12:2.)

This clarifies when the siege will take place: it will be in a time when Judah and Jerusalem are united. This has only recently occurred. From the time that the Romans scattered the Jews (approximately 70 A.D.) until Jerusalem again became part of the State of Israel (in 1967), there was no "Judah and Jerusalem." Furthermore, Jerusalem has not been "a cup of trembling" (that is, Israel has not caused anyone to tremble with fear) since the days of King David, when the House of Israel was at its zenith. However, that has all changed in our time. Since its creation in 1948, the tiny nation of modern Israel has taken territory and inflicted tremendous losses on any aggressor nation choosing to test its prowess in battle. All this seems to fulfill the requirements of Zechariah's prophecy, and indicate that the siege to which Zechariah referred may be in the very near future.

Ezekiel also may have beheld in vision the onset of the same siege spoken of by Zechariah.

> And thou shalt come up against my people of Israel, as a cloud to cover the land; it shall be in the latter days, and I will bring thee against my land, that the heathen may know me, when I shall be sanctified in thee, O Gog, before their eyes. (Ezekiel 38:16.)

(As we shall see later in this Chapter, both Ezekiel and Zechariah declared that the Savior would appear at that time, which seems to verify that they were both describing the same event.)

The might of the Lord

Zechariah described the coming battle in vivid terms:

> For I will gather all nations against Jerusalem to battle; and the city shall be taken, and the houses rifled, and the women ravished; and half the city shall go forth into captivity, and the residue of the people shall not be cut off from the city. Then shall the Lord go forth and fight against those nations, as when he fought in the day of battle. (Zechariah 14:3.)

It is important to note from Zechariah's account that the power of the Lord—not the might of the Jews—is to be thrown against the enemy. But that statement raises some questions. If it is not the armed might of the Israelite Army that overcomes the invaders, how will the Lord destroy the enemies of the Jews? And when was "the day of battle" to which Zechariah referred, when the Lord fought Israel's enemies—apparently a past event which he said will be like the future deliverance?

That reference is an extremely important clue to the nature of the future events Zechariah and Ezekiel related, because the earliest reference to the Lord doing battle for the

Israelites occurs in Exodus. Moses sang a song of praise to the Lord for delivering the children of Israel from the Egyptians. (Exodus 15:3.) That Prophet also recounted the many marvelous manifestations of the Lord's power in saving the Israelites from destruction. Zechariah's reference to Exodus events in order to explain future events further strengthens the thesis that these prophesied events can be explained by understanding the correct nature of past events—*especially those of the Exodus.*

In addition, the Book of Joshua describes what happened on the day the sun and the moon stood still, when the Lord again fought for Israel, and then comments: "And there was no day like that before it or after it . . . for the Lord fought for Israel." (Joshua 10:14.) In the Book of Jasher (an extra-Biblical source mentioned in the Book of Joshua), Joshua sings a song of praise to the Lord, saying that God is "mighty in battle," after destroying the Amorites by "casting down great stones upon them" and causing the sun and the moon to stand still so that Joshua and his army could finish the battle. (*The Book of Jasher,* pp. 217-18.)

Exodus miracles to be seen again

Those occasions (as indicated by Moses and Joshua) when the Lord did battle for the Israelites are both cited by Velikovsky as times of planetary encounters. Therefore, it appears that Zechariah compared the Israelites' deliverance during the Exodus and Joshua's great battle day to the deliverance of the Jews and Jerusalem in the last days. This strongly suggests that the deliverance during the siege in the last days will also be occasioned by a celestial catastrophe similar to those that occurred in the days of Moses and Joshua. Zechariah and Ezekiel both emphasized that the deliverance of the Jews and the City of Jerusalem in the future battle could occur in a manner similar to those of the Exodus and the long day of Joshua.

Note the phenomena Zechariah mentions as he describes how the Lord will save Jerusalem:

> And his feet shall stand in that day upon the mount of Olives, which is before Jerusalem on the east, and the mount of Olives shall cleave in the midst thereof toward the east and toward the west, and there shall be a very great valley; and half of the mountain shall remove toward the north, and half of it toward the south. (Zechariah 14:4.)

Next, Zechariah wrote that the city's inhabitants would "flee, like as ye fled from before the earthquake in the days of Uzziah, King of Judah." (Zechariah 14:5.) This is another invaluable clue to the nature of the coming event. Again, the Prophet is referring us to past events so we can understand better what will happen during the future battle for Jerusalem. Upon referring to history, another fascinating correlation is discovered: a great earthquake occurred in Jerusalem and the surrounding area in the days of Uzziah, as the Prophet Amos had predicted earlier. But it apparently was no ordinary earthquake. Velikovsky cited that event as one effect of the first Earth-Mars encounter, approximately 747 B.C., during Uzziah's reign.

Isaiah, who likely began his ministry on the very day of the "commotion" during the days of Uzziah, left this account of the conditions in and around Jerusalem at that time: "Your country is desolate, your cities are burned with fire . . . . Except the Lord of hosts had left unto us a very small remnant, we should have been as Sodom, and we should have been like unto Gomorrah." (Isaiah 1:7, 9.) His account leaves little doubt that most of Jerusalem's inhabitants and much of the city had been destroyed in Uzziah's time. In earlier times, Sodom and Gomorrah had been totally consumed. Isaiah was comparing those two destroyed cities with Jerusalem after the "commotion" in Uzziah's time, to illustrate the extensive damage. Only a "very small remnant" remained. Therefore, by mentioning the catastrophe Isaiah wrote about, during Uzziah's reign, the Prophet Zechariah is pointing out what the destructions in Jerusalem will be like during that great battle in the last days.

Can it be mere coincidence that both Ezekiel and Zechariah referred to ancient events in order to clarify their descriptions of the future siege of Jerusalem? Knowing the similar nature of both past and future events, the Prophets naturally drew on appropriate parallels, providing excellent keys for understanding their prophecies. All that was needed was a correct understanding of the nature of past catastrophic events in order to comprehend the exact meaning the Prophets intended to convey.

Cosmic cataclysm

The time of Jerusalem's siege will see a re-enactment of the cosmic near-collision scenario in which a tremendous earthshock will shake and rend our entire planet. The Mount of Olives will split north and south, creating a new valley running east toward the Dead Sea. Speaking of this earthquake, Ezekiel said there would be "a great shaking in the land of Israel," and he declared: "All the men that are upon the face of the earth, shall shake at [the Lord's] presence, and the mountains shall be thrown down, and the steep places shall fall, and every wall shall fall to the ground." (Ezekiel 38:20.)

Day and night confused

Zechariah continued to describe what would happen when Jerusalem was delivered:

> And it shall come to pass in that day, that the light shall not be clear, nor dark:

> But it shall be one day which shall be known to the Lord, not day, nor night: but it shall come to pass, that at evening time it shall be light. (Zechariah 14:6, 7.)

This clearly refers to the confusion of day and night that attends such events, as we have learned. Zechariah's description adds to the evidence that this event will be caused by a cosmic catastrophe. The prolongation of day and night is reminiscent of the Exodus and the long day of Joshua, when the Earth's rotation was disrupted. This comment by Zechariah also may refer to the darkened atmosphere that occurs during such catastrophes.

A new river

There is yet more detail in Zechariah's vision:

> It shall be in that day, that living waters shall go out from Jerusalem; half of them toward the former sea, and half of them toward the hinder sea: in summer and in winter shall it be. (Zechariah 14:8.)

Such a spring of "living waters" might naturally result from the shifting of the Earth's crust during an upheaval. Ezekiel confirmed that water would begin to flow out of Jerusalem toward the Dead Sea, and he went to great lengths to point out that it would be a large river. (See Ezekiel 47:5.)

Zechariah then described how the topography of the region surrounding Jerusalem will be altered by the earthquake:

> All the land shall be turned as a plain from Geba to Rimmon south of Jerusalem: and it shall be lifted up, and inhabited in her place, from Benjamin's gate unto the place of the first gate, unto

> the corner gate, and from the tower of Haenaneel unto the king's winepresses. (Zechariah 14:10.)

A blast from heaven

Having seen what is to befall Jerusalem, we may now consider how the planetary encounter will affect the armies surrounding and occupying that holy city. Wrote Zechariah:

> And this shall be the plague wherewith the Lord will smite all the people that have fought against Jerusalem; their flesh shall consume away while they stand upon their feet, and their eyes shall consume away in their holes, and their tongue shall consume away in their mouth. (Zechariah 14:12.)

Recall that a "blast from heaven" destroyed the Assyrian Army of Sennacherib when it surrounded Jerusalem in the days of Hezekiah, King of Judah. (Isaiah 37:7.) (This is another event from the past, cited by Velikovsky as a time of marauding planets, that may be instructive.)

It is interesting to note that Talmudic and Midrashic sources comment on how those Assyrian soldiers died, saying that "their souls were burnt, though their garments remained intact." (See *Worlds in Collision,* p. 237.) This may be a manifestation of the interplanetary lightning which Velikovsky described as one strange and powerful phenomenon of the Exodus. Anyone near to where such a tremendous bolt of lightning touched down would certainly be "consumed." It may well be that Zechariah is referring to the same phenomenon when he says of Jerusalem's future besiegers that "their flesh shall consume away while they stand upon their feet." However, the description by Zechariah could have also referred to the effects of the great plague, another of the dire manifestations of a near-collision, which will destroy so much of mankind in the last days.

Familiar signs

The Prophet Ezekiel added more about that great battle to come in the last days:

> I [the Lord] will plead against him [the army that surrounds Jerusalem] with pestilence and with blood; and I will rain upon him, and his bands, and upon the many people that are with him, an overflowing rain, and [of] great hailstones, fire, and brimstone." (Ezekiel 38:22.)

(Notice all the references to cataclysmic manifestations—pestilence, blood, meteorites, fire, ash and so on. This is convincing evidence that this future event will be occasioned by a cosmic catastrophe.)

Many will die

Ezekiel also said that the Lord will leave only one-sixth of the besieging army alive. (Ezekiel 39:2.) Zechariah observed that people living in Jerusalem will die during these calamities:

> And it shall come to pass, that in all the land, saith the Lord, two parts therein shall be cut off and die; but the third shall be left therein.
>
> And I will bring the third part through the fire, and will refine them as silver is refined, and will try them as gold is tried: they shall call on my name, and I will hear them . . . (Zechariah 13:8, 9.)

Make no mistake—the destructions to take place in and around Jerusalem will occur worldwide. The ratio of 2 deaths to 1 survivor may also apply worldwide, for Zechariah wrote that the devastation will occur "in all the land."

The Savior appears

Both Zechariah and Ezekiel clearly indicated that the Savior will appear to survivors of the destruction. The Prophet wrote, "And his feet shall stand in that day upon the Mount of Olives, which is before Jerusalem on the east . . ." (Zechariah 14:4.) This will present an awkward situation for the Jews. They will realize from the prints in his hands and his feet that this is not only their promised Messiah, but also Jesus of Nazareth—crucified by their forefathers in the meridian of time.

> And one shall say unto him, What are these wounds in thine hands? Then he shall answer, Those with which I was wounded in the house of my friends. (Zechariah 13:6.)
>
> . . . and they shall look upon me whom they have pierced, and they shall mourn for him, as one mourneth for his only son, and shall be in bitterness for him, as one that is in bitterness for his firstborn. In that day shall there be a great mourning in Jerusalem . . . (Zechariah 12:10.)

Modern revelation has given us the conversation that will take place between the Savior and the Jews at the time of his appearance to them.

> And then shall the Jews look upon me and say: What are these wounds in thine hands and in thy feet?
>
> Then shall they know that I am the Lord; for I will say unto them: These wounds are the wounds with which I was wounded in the house of my friends. I am he who was lifted up. I am Jesus that was crucified. I am the Son of God.
>
> And then shall they weep because of their

> iniquities; then shall they lament because they persecuted their king. (Doctrine & Covenants 45:51-53.)

One cannot help but be moved by the poignancy of that bittersweet moment.

Apparently the Jews are not the only ones who will begin, once again, to believe in Jesus Christ. Zechariah indicated that all those in and around Jerusalem who survive the cataclysms of that future time will be converted.

> And it shall come to pass that every one that is left of all the nations which came against Jerusalem shall even go up from year to year to worship the King, the Lord of Hosts, and to keep the feast of tabernacles. (Zechariah 14:16.)

One is struck with the similarity of the events written in the Book of Mormon that tell of the Savior's appearance to the survivors of the great cataclysm after his death and resurrection. The parallel between his appearance to the Nephites nearly 2000 years ago, and his coming appearance to the surviving Jews in Jerusalem, is a stunning one. The sequence of events as related in the Book of Mormon account may provide a blueprint of the sequence of events to yet occur in Jerusalem.

The temple problem

Ezekiel and Zechariah provided us with a fairly detailed account of the events surrounding the Savior's appearance in Jerusalem in the last days. Ezekiel was shown in vision that a river would flow out of Jerusalem from beneath the mount on which a new temple would be built. (Ezekiel 47:1.) Temples built in Jerusalem have always been constructed on Mount Moriah, the location dictated by strong religious tradition as the site of the Temple of God. At present there is no Jewish temple on Mount Moriah. Instead, a Moslem mosque called

"Mosque of Omar," or "Dome of the Rock" (second only to Mecca in its importance in Islam) occupies the spot the Jews believe was their temple's location in ancient times. Orthodox Jews in Jerusalem would like to rebuild the temple at the location indicated by their traditions; but in order to do so, it first would be necessary to demolish the Moslem mosque. That currently is an unthinkable alternative, for all Islam would rise up and declare a holy war, or *jihad,* on Israel. So, at present, the idea of rebuilding the temple is out of the question.

The solution

How do the Jews expect to overcome the obstacle the Dome of the Rock represents? If an orthodox Jew living in Jerusalem were to speak candidly, he would likely say that the matter is in the hands of God, and the visions of Ezekiel and Zechariah seem to bear out that assessment.

A summary of Jewish prophecies indicates that:

When the cataclysms begin, the entire Earth will be convulsed by an earthquake which will split the Mount of Olives east of Mount Moriah and open a fissure or valley eastward toward the Dead Sea.

Every building in Jerusalem—including the Dome of the Rock—will be leveled by the quake that splits the Mount of Olives and radically alters the topography of that region.

A river will erupt from beneath Mount Moriah and flow eastward through the newly-formed fissure where the Mount of Olives once stood, and from there on to the Dead Sea.

The survivors, converted by the appearance of the Savior, then will proceed to rebuild Jerusalem and the Temple of God on Mount Moriah.

So not only will the cataclysm mean the destruction of

the Moslems' great mosque, but it also will see the Moslems, as well as the Jews, converted to a belief in Jesus Christ. (This is plainly in keeping with Hebrew and Arabic tradition since the Moslem Arabs also correctly claim to be descendants of Abraham and heirs to many of the same blessings as the House of Israel.)

Not the Second Coming

Many have supposed that the Savior's appearance at the time of the siege of Jerusalem is his coming in glory; however, this is not the case. Much of the evidence gathered from prophecy indicates that his appearance on the Mount of Olives is a preliminary appearance.

Latter-day Saints have been instructed that the Lord will appear on several occasions prior to his glorious Second Coming. Joseph Smith taught that one occasion when the Savior will appear will be during the future Council of Adam-ondi-Ahman. We know that he has already appeared in the Kirtland Temple. The Savior's appearance on the Mount of Olives is rather a limited visitation to prepare his people for the last great event.

In the Book of Revelation, John saw three separate "woes." As discussed in Chapter 7, John's term "woe" referred to the calamities seen to accompany a near collision of Earth and another equally large body. The last of these three disastrous encounters immediately precedes the Savior's glorious Second Coming. However, the second woe seems to occur at a time when Jerusalem will be under siege—apparently the same siege to which Zechariah and Ezekiel referred. Though John says nothing of the Savior appearing in Jerusalem, the implication is strongly in favor of that conclusion. All the scriptural accounts dealing with that event agree in these two particulars: it is a time when the Jews and Jerusalem are under siege, and the destruction that occurs at that time is of a catastrophic nature.

This sequence of events correctly fits those outlined by the Prophet Joseph Smith:

> Judah must return, Jerusalem must be rebuilt, and the temple, and water come out from under the temple, and the waters of the Dead Sea be healed. It will take some time to rebuild the walls of the city and the temple, &c.; and all this must be done *before the Son of Man will make His appearance* [the Second Coming]. There will be wars and rumors of wars, signs in the heavens above and on the earth beneath, the sun turned into darkness and the moon to blood, earthquakes in divers places, the seas heaving beyond their bounds; then will appear one grand sign of the Son of Man in heaven. (*Documentary History of the Church*, 5:337; italics added.)

Of course, after such a cataclysmic event, Jerusalem will have to be completely rebuilt, and it will then be possible for the Temple of God to be built in place of the present mosque. Furthermore, rebuilding Jerusalem will be quite an undertaking for the survivors, and will likely demand several years of work—even with help from other nations.

Much more detail in the prophecies which deal with the siege of Jerusalem will not be covered here for lack of space. Many fascinating things have been predicted, and Latter-day Saints should become familiar with them. This analysis of the siege of Jerusalem has provided yet another example of how correct concepts of the past assist in understanding events prophesied in the future.

". . . and as the words have gone forth out of my mouth even so shall they be fulfilled, that the first shall be last, and that the last shall be first in all things whatsoever I have created by the word of my power . . . ."

Doctrine & Covenants 29:30

9

When They Return

There is yet one other application of the thesis put forth by Dr. Immanuel Velikovsky which, in light of statements made by or attributed to latter-day Prophets, seems to have remarkable implications.

Taught by the Prophet

Elder Joseph Young, of the First Council of the Seventy, and brother to Brigham Young, claimed to have learned the following from Joseph Smith himself:

> . . . Enoch and his people, with the city which they occupied, and the foundations on which it stood, with a large piece of earth immediately connected with the foundations and the city, had assumed an aerial position within the limits of our solar system. . . .

> Joseph Smith said, on another occasion, in the hearing of some of the saints still surviving [in about 1880] that the City of Enoch would again take its place in the identical spot from which it had been detached, now forming the chasm of the earth, filled with water, called the Gulf of Mexico. (*Enoch And His City*, pp. 11, 12, as quoted in *L.D.S. Reference Encyclopedia*, p. 121.)

It is interesting to note that none of the Latter-day Saints alive when Joseph Young published his book appear to have ever challenged that statement. As we reflect on this assertion that the City of Enoch would return and fill the Gulf of Mexico, recall that John wrote of a large body falling into the sea at some future time. "And the second angel sounded, and as it were a great mountain burning with fire was cast into the sea. . . ." (Revelation 8:8.) Is this parallel a coincidence, or evidence in scripture of this event?

Falling stars

Parley P. Pratt answered the following question:

> How can the stars fall from heaven to the earth, when they (as far as we know) are much larger than the earth? We are nowhere given to understand that all the stars will fall or even many of them; but only 'as a fig tree casteth her untimely figs when she is shaken with a mighty wind.' The stars that will fall to earth are fragments, which have been broken off from the earth from time to time in the mighty convulsions of nature . . . . These all must be restored again at the 'times of restitution of all things.' . . . When these fragments, (some of which are vastly larger than the present earth) are brought back and joined to this earth, it will cause a convulsion of all nature . . . the mountains will flow

> down, the valleys rise, the sea retire to its own place, the islands and continents will be removed, and the earth be rolled together as a scroll . . . (*Millennial Star*, I:258.)

Elder Pratt was speaking of events yet future, and yet his statement contains the very elements outlined by Velikovsky as indicative of a near collision of planets. Recall that the Apostle John wrote, "And the fifth angel sounded, and I saw a star fall from heaven unto the earth. . . . (Revelation 9:1.) Could it be that he was referring to the body that will return to earth bearing its terrestrial inhabitants?

A planetary reunion

Patriarch Homer M. Brown declared that his grandfather was taught much the same thing by Joseph Smith. (*The Last Days*, pp. 212-215.) He claimed that the Prophet taught him that this collision would not be head-on, but rather the overtaking of one planet by the other—both moving in the same direction (what we of the space age would call a docking maneuver). He also was taught that the water receding to the north countries at the appearance of the Lost Tribes (Doctrine & Covenants 133:23) would be a consequence of the slowed rotation of the earth, caused by the other body.

The prophet understood

From the Journal of Wandle Mace, who recorded many events and teachings of the life of Joseph Smith, comes an account of an address given by Joseph Smith in the Jack Oak Grove in Nauvoo, during which Joseph commented on the Lost Tribes. Mace stated that he taught that parts of the Earth had been "taken away" in the past to form abodes for various groups of people, and that those parts would return in the future, bearing their inhabitants. He quoted the Prophet as saying that "the earth will be restored as at the beginning, and the last taken away will be the first to return, for the last shall be first and the first shall be last in all things." (*Journal of Wandle Mace*, p. 47.)

Mace further quoted Joseph Smith's illustration of the effect of the returning body on the Earth and its inhabitants:

> Some of you brethren have been coming up the river on a Steamboat, and while seated at the table, the Steamboat [ran] against a snag which upset the table and scattered the dishes; so it will be when these portions of earth return. It will make the earth reel to and fro like a drunken man. (*Journal of Wandle Mace*, p. 47.)

There are many such second hand sources which attribute this teaching to Joseph Smith, and while the validity of such statements can be challenged to one degree or another, most are fairly credible.

Do not overlook the incredible

It may be worthy of note to point out that the diaries and writings of early Church members occasionally attribute statements that seem incredible to one or another of the early Church leaders. Just as with the statements of the ancient Prophets, though the explanation for such extravagant statements may not be readily apparent, we should perhaps at least consider them since they may well prove to be accurate. It will not do to simply disregard or dismiss such extraordinary statements because they seem extravagant to our current understandings.

Earth fragments

One of the most credible sources noting this teaching was Eliza R. Snow Smith, a wife of Joseph Smith. Noted in L.D.S. History as an extremely capable, intelligent and enlightened woman, she learned the gospel and gained much of her wisdom from the Prophet himself.

In a poem about the Earth (which later was employed as the lyrics for an LDS hymn), she, like Elder Pratt, asserted that portions of the Earth had been torn away in the past to serve as a dwelling place for the City of Enoch and the Lost Tribes:

> Thou, earth, wast once a glorious sphere
> Of Noble magnitude,
> And didst with majesty appear,
> Among the worlds of God.
>
> But thy dimensions have been torn
> Asunder, piece by piece,
> And each dismembered fragment borne
> Abroad to distant space.
>
> When Enoch could no longer stay
> Amid corruption here,
> Part of thyself was borne away
> To form another sphere.
>
> And when the Lord saw fit to hide
> The "ten lost tribes" away,
> Thou earth, wast severed to provide
> The orb on which they stay.
>
> And thus, from time to time, thy size
> Has been diminished til
> Thou seemest the law of sacrifice
> Created to fulfill.
>
> A "restitution" yet must come,
> That will to thee restore,
> By that grand law of worlds, thy sum
> Of matter heretofore.
> (Extracts from Hymn 322, *Sacred Hymns and Spiritual Songs*, pp. 386-87.)

While there may be numerous theories concerning the whereabouts of the Lost Tribes of Israel or the City of Enoch, the relationship between these teachings attributed to Joseph Smith and Velikovsky's thesis of catastrophism and interplanetary encounters is very intriguing. If Velikovsky's theory is applicable to the return of the City of Enoch and the Lost Tribes, perhaps a brief examination of prophetic accounts of their return would be worthwhile.

Cataclysmic return

The Doctrine & Covenants contains the following concerning the return of the Lost Tribes:

> And he shall utter his voice out of Zion, and he shall speak from Jerusalem, and his voice shall be heard among all people;
>
> And it shall be a voice as the voice of many waters, and as the voice of a great thunder, which shall break down the mountains, and the valleys shall not be found.
>
> He shall command the great deep, and it shall be driven back into the north countries, and the islands shall become one land: [i.e., the reuniting of the continents];
>
> And the land of Jerusalem and the land of Zion shall be turned back into their own place, and the earth shall be like as it was in the days before it was divided [i.e., before various portions were divided off planet earth].
>
> And they who are in the north countries shall come in remembrance before the Lord; and their Prophets shall hear his voice, and shall no longer stay themselves; and they shall smite the rocks, and the ice shall flow down at their presence.

> And an highway shall be cast up in the midst of the great deep. . . .
>
> And in the barren deserts there shall come forth pools of living water; and the parched ground shall no longer be a thirsty land. . . .
>
> And the boundaries of the everlasting hills shall tremble at their presence. (Doctrine & Covenants 133:21-24, 26-27, 29, 31.)

It is evident that this event, or series of events, will be catastrophic in nature. For the ice in the north countries to melt or "flow down" at their arrival, tremendous heat would have to be generated. We have seen how such heat could be created in a planetary encounter.

"An highway" that is "cast up" in the ocean or "great deep," would be caused by a major upheaval of the Earth's crust. Such movement would naturally cause "the everlasting hills" to be shaken by an earthquake, i.e., "tremble at their presence." Shifting strata would cause new springs of water to erupt or change river courses so that the "parched ground [would] no longer be a thirsty land."

Tidal recession

One of the manifestations of a near-collision of the Earth and an intruder would probably be that of slowed rotation. The effect of this upon the water in the oceans would be identical to the effect cited by the Lord in the preceding verses. As the Earth's rotation slowed, the centrifugal force which causes the Earth's equatorial bulge would be lessened, and the oceans would shift away from the equator, toward the poles.

There can be little doubt that the return of the Lost Tribes would occasion disruptions of global magnitude, and the accompanying natural manifestations mentioned are some of the same as those we find in the list of destructions in a planetary close encounter.

Elder Orson Pratt cites the appearance of "the glory of God overshadowing their camp by day in the form of a cloud, and a pillar by night" just prior to the coming of the Lost Tribes (*Journal of Discourses,* 25:34)—the same allusion to a pillar of fire and a cloud that appears in Exodus. The implication here is that at this future encounter, the body will not simply pass close by, but will join with the Earth in its orbit around the Sun.

Restoration of all things

Nearly all those who claimed to have heard Joseph Smith teach these things, told of his teaching it in conjunction with the doctrine of restoration. We know from scripture that this dispensation will see the restoration of all things. We correctly assume that that means the restoration of the gospel, the Priesthood, and the ordinances for salvation and exaltation, along with the organization of the true Church. Yet, perhaps the doctrine of restoration extends to other things as well. Apparently, Joseph Smith taught that it also meant a physical restoration of the Earth to its original form and condition. Recall that Elder Pratt taught that some of the fragments which are to return and *join again with the Earth* "are vastly larger than the present earth." (*Millennial Star,* I:258.) It would seem that our Earth once appeared very differently than it does today.

Velikovsky's thesis seems to provide an explanation as to how the prophecies of the return of the Lost Tribes and the City of Enoch might be fulfilled. Until now, these statements were largely ignored because the mechanism for such fantastic events was not understood. Yet the concept apparently taught by Joseph Smith seems to be reinforced by almost all of the prophecy which we have concerning the latter days.

"Blessed is he that readeth, and they that hear the words of of this prophecy, and keep those things which are written therein: for the time is at hand."

Revelation 1:3

10

That We May Better Understand

It is recorded that as the Savior sat upon the Mount of Olives, his disciples came to inquire of him, saying:

> Tell us when shall these things be which thou hast said concerning the destruction of the temple, and the Jews; and what is the sign of thy coming, and the end of the world, or the destruction of the wicked, which is the end of the world? (Joseph Smith-Matthew 1:4.)

These are not unlike the questions we would ask of prophetic accounts which deal with the Savior's Second Coming. To us, as well as to those disciples who besought him anciently, the Savior responded with an injunction to "take heed that no man deceive you . . . he that remaineth steadfast and is not overcome, the same shall be saved." (Joseph Smith-Matthew 1:5, 11.) Perhaps it would be useful for us to examine the Savior's answer in light of our current approach.

Know the truth

In order to better understand the Savior's prophecies recorded in Joseph Smith's translation of Matthew, it would be helpful to refer first to the vision of the Apostle John in Revelation. In the Sixth Chapter of Revelation, John recorded the judgments which are to befall the world before the Savior comes again. After the cataclysms, John said, the Earth's inhabitants

> . . . hid themselves in the dens and in the rocks of the mountains;
>
> And said to the mountains and rocks, Fall on us, and hide us from the face of him that sitteth on the throne, and from the wrath of the Lamb: for the great day of his wrath is come; and who shall be able to stand? (Revelation 6:15-17.)

In other words, when the sun becomes black as sackcloth of hair, and the moon is turned red as though it were blood, and the stars fall, and every mountain and island is moved from its place, Earth's inhabitants will assume that the day of the Lord's coming is upon them. But the Savior's warning to his apostles, and to us, is clear: they would be terribly mistaken to believe that the Savior had come. If prophecy told us no more than that his appearance would be preceded by destruction, we too might be apt to expect the Savior's coming to follow the initial devastation. We might be wont to harken to those who declare: "'Behold, he is in the desert . . . Behold, he is in the secret chambers.'" But the Savior's caution to his disciples, and to us, is clear: *"believe it not!"*

Only the beginning of the end

> All things which have befallen them are only the beginning of the sorrows which shall come upon them.
>
> . . . see that ye be not troubled, for all I have told you must come to pass; *but the end is not yet.* (Joseph Smith-Matthew 1:25, 19, 23.)

The Lord seems to emphasize the point that when the first signs appear we should not mistake them as signifying the end, but only the beginning of the end. The first signs will not be the signs of his actual coming, so he warns us against expecting to see him immediately after they occur. Instead, the first signs are only the "beginning of sorrows."

A type and a kind

The Book of Mormon provides us with an example of the type of persecution to which the faithful may be subjected in the last days. The Prophet Samuel had foretold the signs of Christ's birth, and the righteous had looked forward to that day:

> But there were some who began to say that the time was past for the words to be fulfilled, which were spoken by Samuel, the Lamanite.
>
> And they began to rejoice over their brethren, saying: Behold the time is past, and the words of Samuel are not fulfilled; therefore, your joy and your faith concerning this thing hath been vain.
>
> And it came to pass that they did make a great uproar throughout the land; and the people who believed began to be very sorrowful, lest by any means those things which had been spoken might not come to pass.

> Now it came to pass that there was a day set apart by the unbelievers, that all those who believed in those traditions should be put to death except the sign should come to pass, which had been given by Samuel the prophet. (3 Nephi 1:5-7, 9.)

Thus, it may be in the last days that people will expect to see the Savior coming in glory after the first great destruction or "woe," as John termed it. When he does not appear, some disillusioned individuals (like those spoken of in the above quote) may begin to say, "'Where is the promise of his coming?'" (II Peter 3:4.) After all, the signs of his coming will have been made catastrophically apparent. Anyone having a passing acquaintance with prophecy and the signs of the last days would naturally expect the Savior to appear subsequent to the prophesied signs and destructions. Other individuals, likely, will seek to deceive the faithful by saying that the Savior has come, but that he is only to be found in a special place, saying, "'Behold, he is in the desert,'" or "'Behold, he is in the secret chambers.'" (Joseph Smith-Matthew 1:25.) Still others will represent themselves as "false Christs, and false prophets." These, the Lord warns, "shall shew great signs and wonders" in order to mislead the faithful. Again, his admonition to us is emphatic: "believe it not!" (Joseph Smith-Matthew 1:22, 25.)

Endure to the end

And why should there be so much confusion as to when the Savior will again appear? Simply because there will be several occasions when the same or very similar signs will be made manifest prior to the Savior's advent. If we are to take literally the three woes John prophesied in Revelation, then the faithful may have to endure two or more separate occasions of destruction without the accompanying visitation of the Savior. It may happen that he will appear at a time when a good many have lost hope, when the world persecutes the faithful for waiting patiently upon the Lord. Of these possibilities, Jesus warned his disciples:

> And what I say unto one, I say unto all men; watch therefore, for you know not what hour your Lord doth come.
>
> But know this, if the good man of the house had known in what watch the thief would come, he would have watched, and would not have suffered his house to have been broken up, but would have been ready.
>
> Therefore be ye also ready, for in such an hour as ye think not, the Son of man cometh. (Joseph Smith-Matthew 1:46-48.)

Peter also wrote of the last days, using the same metaphor as the Savior did. He explained that the Lord's coming would be

> . . . as a thief in the night; in the which the heavens shall pass away with a great noise, and the elements shall melt with fervent heat, and the earth also and the works that are therein shall be burned up.
>
> Wherefore, beloved, seeing that ye look for such things, be diligent that ye may be found of him in peace, without spot, and blameless.
>
> Ye therefore, beloved, seeing ye know these things before, beware lest ye also, being led away with the error of the wicked, fall from your own steadfastness. But grow in grace, and in the knowledge of our Lord and Saviour Jesus Christ. (II Peter 3:10, 11, 14, 17, 18.)

The lesson taught by the Savior to the Twelve

concerning the events of the last days is likely the most important we could learn from prophecy. Instead of simply enumerating the multitude of signs, wonders and catastrophes that will occur, he chose to warn them, and us, of anticipating his return before it will actually occur.

A hope for survival

A study of the nature of the cataclysms which are to occur prior to the Second Coming may also cause us some anxiety for our survival and the survival of our loved ones. Nevertheless, the Lord has declared: "If ye are prepared ye shall not fear." (Doctrine & Covenants 38:30.) When the Prophets (ancient and modern) speak of preparation for the last days, they speak in terms of personal righteousness. The message of the scriptures, notwithstanding the calamities which they enumerate, is a positive message—a message of hope. Prophecy is given that we might know what to expect, and we have been given the gospel and the living Prophets that we might know how to prepare for that which is to come. This spiritual preparation—the opportunity for salvation—is available to all men and women.

President Wilford Woodruff made this statement:

> Can you tell me where the people are who will be shielded and protected from these great calamities and judgments which are now at our doors? I'll tell you. The priesthood of God who honor their priesthood, and who are worthy of their blessings are the only ones who shall have this safety and protection. They are the only mortal beings. No other people have a right to be shielded from these judgments. They are at our very doors; not even this people will escape them entirely. They will come down like the judgments of Sodom and Gomorrah. And none but the priesthood will be

> safe from their fury. (*Young Women's Journal*, Vol. 5, No. 11.)

Elder Bruce R. McConkie, speaking of "The Coming Tests and Trials and Glory," provided what may be the most pertinent and poignant statement of this subject:

> If we, as a people, keep the commandments of God; if we take the side of the Church on all issues, both religious and political; if we take the Holy Spirit for our guide; if we give heed to the words of the Apostles and Prophets who minister among us—then, from an eternal standpoint, all things will work together for our good. Our view of the future shall be undimmed, and whether in life or in death, we shall see our blessed Lord return to reign on earth. We shall see the New Jerusalem coming down from God in heaven to join with the Holy City we have built. We shall mingle with those of Enoch's city while together we worship and serve the Lord forever. (*Ensign*, May 1980, p. 73.)

We should not confuse survival with salvation. The gospel teaches us that spiritual salvation, not temporal salvation, is the object of our existence. Eternity beckons to us all. No matter the event that takes us, we will be blessed according to our righteousness. The question is not so much whether we will live, as how.

Israel, ancient and modern

Finally, as we consider that we are modern Israel, and that the prophecies concerning the times in which we live may be understood in terms of the Exodus and the ancient Israelites, we should consider the profound implications of the analogy.

Throughout the scriptures, the same story is told time and time again. The names and the places change, but the essence of the story never does. It is this: a prophet is called of God. His missionary efforts convert many righteous individuals who are ultimately forced to flee to the "wilderness" in order to escape the persecution and wickedness of the world. Eventually they are delivered from the wilderness to a "promised land."

Enoch, Abraham, Moses, Lehi, Alma and countless others have been the principal actors in such a drama. Enoch, for example, was called of God to preach repentance. He was a most successful missionary, but the world persecuted the righteous which he had found, and so the Lord gave Enoch great power over his enemies and allowed him to gather the Saints of his day to a city of their own which they called Zion. Jared was instructed to take those who would believe on his word and flee the land of the Tower of Babel. He was guided by the word and power of the Lord insomuch that he was able to do great and marvelous things for the benefit of his followers. Through their fathfulness they were led to a land of promise. Abraham also was called to gather a righteous few and flee to a new land which the Lord gave not only to him, but to his posterity forever. Thus the story has been retold in the lives of many, many Prophets and Saints.

And so it is today

The principal actors in this dispensation are the Prophet Joseph Smith and the Latter-day Saints. There was an initial gathering in the days of Joseph Smith, but the Church was forced to flee to the wilderness because of persecution. Today the gathering still continues; but now it is to the stakes of Zion, not to a city of Zion. We, as a people, are still at the "sojourn in the wilderness" stage. The next stage is that of a deliverance to "the promised land."

This parallel was strongly reinforced by the Lord when he said through Joseph Smith:

> Behold, I say unto you, the redemption of Zion must needs come by power;
>
> Therefore, I will raise up unto my people a man, who shall lead them like as Moses led the children of Israel.
>
> For ye are the children of Israel, and of the seed of Abraham, and ye must needs be led out of bondage by power, and with a stretched-out arm.
>
> And as your fathers were led at the first, even so shall the redemption of Zion be. (Doctrine & Covenants 103:15-18.)

This prophecy has not yet been fulfilled. It was not fulfilled in the days of Joseph Smith, for Zion was not then redeemed. Brigham Young certainly fit the picture of a modern Moses; but he, like Joseph Smith, did not lead modern Israel to a redemption of Zion. Rather, he led them into the wilderness.

Saints to be delivered

The inference in the above scripture is unmistakable. In this curious parallelism of the gospel, we are the House of Israel. We will be led out of the wilderness to the promised land. And the message seems clear: the Saints will return to Jackson County (Zion) under circumstances similar to those which the Israelites experienced when they left Egypt and went to their land of promise, for "as your fathers were led at the first, even so shall the redemption of Zion be."

Furthermore, many of the statements of the Brethren concerning the return of the Saints to Jackson County are filled with symbolism of the Exodus. Their statements, taken together with the statements made by the Lord in the Doctrine & Covenants, make a strong case for this idea. For example, Elder Orson Pratt said this:

> We shall go back to Jackson County. . . . When we go back, there will be a very large organization consisting of thousands, and tens of thousands, and they will march forward, the glory of God overshadowing their camp by day in the form of a cloud, and a pillar of flaming fire by night, the Lord's voice being uttered forth before his army. (*Journal of Discourses,* 15:34.)

Even the Lord's characterization of the land of promise contains an unmistakable metaphor:

> And I hold forth and deign to give unto you greater riches, even a land of promise, a land flowing with milk and honey, upon which there shall be no curse when the Lord cometh. (Doctrine & Covenants, 38:18.)

The poetic usage of "a land flowing with milk and honey," came from the effects of manna precipitating in the rivers and streams which then became sweet to the taste and turned a creamy white. Manna was a direct manifestation of past catastrophes. Its use in describing the redemption of Zion adds weight to the conclusion that similar cataclysms will be associated with that great event.

Amid great destruction

It was John Taylor (most likely) who saw the Jackson County area in vision just prior to the return of those who will build New Jerusalem. His description of the appearance of the area implies great destruction will precede their gathering.

> . . . I seemed to be standing on the west bank of

> the Missouri River opposite the City of Independence, but I saw no city. I saw the whole states of Missouri and Illinois and part of Iowa were a complete wilderness with no living human being in them. (Cited in *Wilford Woodruff Journals,* June 15, 1887.)

That statement taken with a statement attributed to Heber C. Kimball by Amanda H. Wilcox makes the concept clear. "[Jackson County] will be swept so clear of its inhabitants that . . . there will not be left so much as a yellow dog to wag his tail." (As quoted in *Prophecy, Key to the Future,* p. 91.)

Today, the area in and around Jackson County, Missouri, is very heavily populated; but the two previous quotes tell us it will be "swept clean." What could cause such devastation?

In the past, great walls of water washed across land masses, sweeping the land clean. Brigham Young said, "The sea will heave itself beyond its bounds, engulphing [sic] mighty cities." (*Journal of Discourses,* 8:123.) Might it be that this same agent will sweep the Missouri river valley clean as part of the destructions prophesied to occur just prior to the Saints' return to the promised land?

All these statements seem to indicate that the return of the Saints to Jackson County will occur during or subsequent to one of the great destructions of the last days. Speaking of the return trek, Joseph F. Smith made this statement:

> They will find the journey back to Jackson County will be as real as when they came out here. Now, mark it. And though you may be led by the power of God 'with a stretched out arm,' it will not be more manifest than the leading the people out here to those that participate in it. They will think there are a great many hardships to endure in

> this manifestation of the power of God. (*Journal of Discourses,* 24:156-57.)

The promise

Modern Israel will then be led by a Prophet. The power of God will deliver them. For the most part, they will be delivered from the destruction they will see all around them. This same destruction that decimates the world will be a deliverance for the Saints. They will be given a promised land, swept clean for their arrival. Modern Israel will be delivered by the power of God and directed by a Prophet, just as ancient Israel was.

Appendix A Biographical Sketch of Dr. Immanuel Velikovsky

Born in Vitebsk, Czarist Russia, June 10, 1895, of Jewish parents, Immanuel Velikovsky had a thorough education in Hebrew and Hebrew writings. Later, the family moved to Moscow where Immanuel studied law, ancient history and medicine. He studied in France and Scotland as well. In 1921 he graduated from the University of Moscow with a degree in medicine.

He relocated to Berlin, where he met and married Elisheva Kramer, an accomplished violinist. While living in Berlin, Dr. Velikovsky compiled *Scripta Universitatis atque Bibliothecae Hierosolymitanarum,* a collection of manuscripts from Jewish scholars around the world. Fellow contributors and editors were Dr. Heinrich Loewe and Dr. Albert Einstein.

In 1923 he and his wife moved to Tel Aviv where Dr. Velikovsky began his medical practice. He also practiced psychoanalysis after studying for a time in Vienna with Dr. Wilhelm Stekel, one of Freud's early disciples.

Dr. Velikovsky was one of the co-founders of the Hebrew University in Jerusalem. His compilation of manuscripts written by Jewish scholars became the source for the courses of study in that newly-born institution. Because of his accomplishments, Dr. Velikovsky was offered the opportunity to become the first President of the Hebrew University—though he graciously declined.

His own research into Freud's work led him to examine closely the hero figure of Moses in Hebrew literature. That led him to a novel reconstruction of ancient history, and his thesis of extraordinary natural disasters in antiquity. In order to continue his research, Velikovsky and his wife immigrated to the United States in 1939.

In 1950 he published his first book, *Worlds in Collision.* He was immediately denounced and vilified by the scientific community. There was some small debate concerning his work, but it was generally dismissed as fanciful and unscientific.

Undaunted, Velikovsky continued to write. He published *Earth in Upheaval,* a book dealing strictly with geologic evidence in support of his thesis. Then he began turning out volume after volume detailing his reconstruction of the chronology of the ancient past.

Since he published his thesis over 30 years ago, the world has entered the space age, and instead of debunking Velikovsky's assertions, modern discoveries have substantiated much of what he claimed would be found when interplanetary exploration became a reality.

Although great debate over his theories still rages in scientific circles, a pattern of consistant success for Dr. Velikovsky's position is beginning to emerge. Prior to scientific confirmation he was able to predict:

High surface temperature of Venus

Anomalous rotation of Venus

Hydrocarbons in the atmosphere of Venus (although this is still hotly debated in scientific circles)

Massive atmosphere of Venus

Emission of radio-noises from Jupiter

Extended magnetosphere of the Earth

Existence and importance of electrical and magnetic forces in the universe

Recent melting and bubbling of the lunar surface

Steep thermal gradient toward the lunar surface

Presence of hydrocarbons and carbides in lunar rock and so on.

Even Einstein, a close friend to Dr. Velikovsky, could not agree with his thesis until word came of the discovery of radio-noises from Jupiter (implying that Jupiter was a dynamic planet—radiating high-energy emissions—rather than a frozen ball in space). This renewed Einstein's interest in Velikovsky's theories because it partially verified his thesis, and Einstein agreed to use his influence in the scientific community to get a fair hearing for Velikovsky's work. Sadly, however, Einstein died a short time later. A copy of *Worlds in Collision* was found open on his desk.

Dr. Velikovsky continued to encourage scientists to test his theories, suggesting areas of inquiry and delineating specific tasks for space probes. As always, he continued to answer his critics forthrightly with documentation and detailed analysis.

Nearly 30 years after publishing *Worlds in Collision,* Immanuel Velikovsky passed away (November 17, 1979).

In 1981, two Voyager spacecrafts, in their fly-by of Jupiter and Saturn, discovered phenomena and conditions that can best be explained by processes originally postulated by Dr. Velikovsky. The name Velikovsky may yet take a place alongside those of Copernicus, Galileo, Newton, et al.

Appendix B
An Overview of the Past

This is an overview of Velikovsky's reconstruction of past changes in the solar system and is reprinted by permission from *Velikovsky Reconsidered,* Stephen L. Talbott, ed. (Doubleday, 1975.)

Collisions and Upheavals

(a fragmentary scenario based on Velikovsky's *Worlds in Collision* and *Earth in Upheaval*)

Immanuel Velikovsky manifests a strong distaste for summaries and popularizations of his books. In the past, many erroneous criticisms of his work have been based upon such popularizations, the critics never having studied his books. And indeed these books, detailed in their arguments and exhaustive in their documentation, do not easily lend themselves to summarization.

Nevertheless, in embarking on a project designed to give the fullest possible coverage of all aspects of Velikovsky's work, the Editors of *Pensee* felt it desirable to reacquaint readers with the flow of events described in his revolutionary reconstruction of recent solar-system history. The evidence, amassed in *Worlds in Collision* and *Earth in Upheaval*, is not presented here, and to those who have not read these works, the events must necessarily appear fanciful and insupportable. This difficulty can be remedied, of course, only by direct reference to the scholarly, evidential texts of Velikovsky himself.

The following brief sketch was prepared entirely without Velikovsky's help. The serious student and scholar should resort to a careful and intensive reading of *Worlds in Collision* and *Earth in Upheaval.*

— — — — — — —

Global cataclysms fundamentally altered the face of our planet more than once in historical times. The terrestrial axis shifted. Earth fled from its established orbit. The magnetic poles reversed.

In great convulsions, the seas emptied onto continents, the planet's crust folded, and volcanoes erupted into mountain chains. Lava flows up to a mile thick spilled out over vast areas of the Earth's surface. Climates changed suddenly, ice settling over lush vegetation, while green meadows and forests were transformed into deserts.

In a few awful moments, civilizations collapsed. Species were exterminated in continental sweeps of mud, rock, and sea. Tidal waves crushed even the largest beasts, tossing their bones into tangled heaps in the valley and rock fissures, preserved beneath mountains of sediment. The mammoths of Siberia were instantly frozen and buried.

Surviving generations recorded these events by every means available: in myths and legends, temples and monuments to the planetary gods, precise charts of the heavens, sacrificial rites, astrological canons, detailed records of planetary movements, and tragic lamentations amid fallen cities and destroyed institutions.

"ALL IS RUIN"

Aware of a link between the circuit of heavenly bodies and the catastrophic ruin of previous generations, the ancients ceaselessly watched the planetary movements. Their traditions recalled that when old epochs dissolved, the new "Age," or "Sun," was marked by different celestial paths. Astronomers and seers diligently watched for any change which might augur approaching destruction and the end of an age.

Prior to the second millennium B.C., ancient Hindu records spoke of four visible planets, excluding Venus, Babylonians, meticulous in their observations, likewise failed to report Venus.

But long before 1500 B.C., Jupiter, for centuries chief among the deities, shattered the serenity of the skies. A brilliant, fiery object, expelled from that planet, entered upon a long elliptical orbit around the Sun. The feared god Jupiter had given birth to the comet and protoplanet Venus.

Terrified, men watched the "bright torch of heaven" as it traversed its elongated orbit, menacing the Earth. Venus, a Chinese astronomical text recalls, spanned the heavens, rivaling the Sun in brightness. "The brilliant light of Venus," records an ancient rabbinical source, "blazes from one end of the cosmos to the other."

The fears of the star watchers were justified. As Venus arched away from one perihelion passage during the middle of the second millennium B.C. (ca. 1450), the Earth approached this intruder, entering first the outer reaches of its cometary tail. A rusty ferruginous dust filtered down upon the globe, imparting a bloody hue to land and sea. The fine pigment chafed human skin, and men were overcome by sickness. Those who sought to drink could not. Rivers stank from the rotting carcasses of fish, and men dug desperately for water uncontaminated by the alien dust. "Plague is throughout the land. Blood is everywhere," bewailed the Egyptian Ipuwer. "Men shrink from tasting, human beings thirst after water . . . . That is our water! That is our happiness! What shall we do in respect thereof? All is ruin."

As recalled by the Babylonians, the blood of the celestial monster Tiamat poured out over the world.

But as the Earth's path carried it ever more deeply into the comet's tail, the rain of particles grew steadily more coarse and perilous. Soon a great hail of gravel pelted the Earth. ". . . there was hail, and fire mingled with the hail, very grievous, such as there was none like it in all the land of Egypt since it became a nation." So recorded the author of Exodus.

Fleeing from the torrent of meteorites, men abandoned their livestock to the holocaust. Fields of grain which fed great cities perished. Cried Ipuwer, "No fruits, no herbs are found. That has perished which yesterday was seen. The land is left to its weariness like the cutting of flax." These things happened, say the Mexican *Annals of Cuauhtitlan*, when the sky "rained, not water, but fire, and red-hot stones."

As our planet plunged still deeper into the comet's tail, hydrocarbon gases enveloped the Earth, exploding in bursts of fire in the sky. Unignited trains of petroleum poured onto the planet, sinking into the surface and floating on the seas. From Siberia to the Caucasus to the Arabian desert, great spills of naphtha burned for years, their billows of smoke lending a dark shroud for human despair.

Our planet was pursuing a near-collision course with the massive comet's head.

Suddenly, caught in an invisible grip, the Earth rocked

violently, its axis tilted. In a single convulsed moment, cities were laid waste, great buildings of stone leveled, and populations decimated.

"The towns are destroyed. Upper Egypt has become waste . . . . All is ruin . . . . The residence is overturned in a minute." Around the world, oceans rushed over mountains and poured into continental basins. Rivers flowed upward. Islands sank into the sea. Displaced strata crashed together, while the shifting Earth generated a global hurricane which destroyed forests and swept away the dwellings of men.

In China, the Emperor Yahou spoke of waters which "overtopped the great heights, threatening the heavens with their floods." Decades of labor were required to drain the valleys of the mainland. Arabia was transformed into a desert by the same paroxysms which may have dropped the legendary Atlantis beneath the ocean west of Gibralter.

With dulled senses, survivors lay in a trance for days, choking in the smoky air.

The tilting axis left a portion of the world in protracted darkness, another in extended day. From the Americas to Europe to the Middle East, records tell of darkness persisting for several days. On the edge of the darkness, the people of Iran witnessed a threefold night and a threefold day. Chinese sources speak of a holocaust during which the Sun did not set for many days, and the land was aflame. Peoples and nations everywhere, uprooted by disaster, wandered from their homelands.

CELESTIAL DRAGON

Led by Moses, the Israelites fled the devastation which brought Egypt's Middle Kingdom to an end. As they rushed toward the Sea of Passage, the glistening comet, in form like a dragon's head, shone through the tempest of dust and smoke.

The night sky glowed brightly as the comet's head and its writhing, serpentine tail exchanged gigantic electrical bolts.

The great battle between the fiery comet's head and the column of smoke—between a light god and a leviathan serpent—was memorialized in primary myths around the Earth. Babylonians told of Marduk striking the dragon Tiamat with bolts of fire. The Egyptians saw Isis and Set in deadly combat. The Hindus described Vishnu battling the "crooked serpent." Zeus, in the account of Apollodorus, struggled with the coiled viper Typhon.

The fugitive Israelites, having reached Pi-ha-khiroth, at the edge of the Red Sea, were pursued by the Pharaoh Taoui-Thom (Typhon). The great sea lay divided before the slave people, its waters lifted by the movement of the Earth and the pull of the comet. Crossing the dry sea bottom, the Israelites escaped from Egypt.

As the comet made its closest approach to Earth, Taoui-Thom moved his armies into the sea bed. But even before the entire band of Israelites had crossed to the far side, a giant electrical bolt flew between the two planets. Instantly, the waters collapsed. The Pharaoh, his soldiers and chariots, and those Israelites who still remained between the divided waters were cast furiously into the air and consumed in a seething whirlpool.

The battle in the sky raged for weeks. A column of smoke by day, a pillar of fire by night, Venus meted destruction to nations large and small. To the Israelites, however, it was an instrument of national salvation.

Through a series of close approaches, the comet's tail, a dreadful shadow of death, cinctured the Earth, wreathing the planet in a thick, gloomy haze that lasted for many years. And so, in darkness, a historical age ended.

Possibly, the human race would have become extinct but for a mysterious, life-giving substance precipitated in the heavy atmosphere—the nourishing "manna" and "ambrosia" described in the ancient records of all peoples. It fell with the morning dew, a sweet, yellowish hoarfrost. It was edible. The ambrosial carbohydrates, possibly derived from Venus's hydrocarbons through bacterial action, filled the atmosphere with a sweet fragrance. Streams flowed with "milk and honey." When heated, this "bread of heaven" dissolved, but when cooled, it precipitated into grains which could be preserved for long periods or ground between stones. Its presence allowed man and beast to survive.

In the new age the Sun rose in the east, where formerly it set. The quarters of the world were displaced. Seasons no longer came in their proper times. "The winter is come as summer, the months are reversed, and the hours are disordered," read an Egyptian papyrus. The Chinese Emperor Yahou sent scholars throughout the land to locate north, east, west, and south and draw up a new calendar. Numerous records tell of the Earth "turning over." An Egyptian inscription from before the tumult says that the Sun "riseth in the west."

While men attempted to determine the times and seasons, Venus continued on its threatening course around the Sun. Under Joshua, the Israelites had entered the Promised Land, and again Venus drew near. It was while the Canaanites fled from before the hand of Joshua in the valley of Beth-horon—some fifty years after the Exodus—that the daughter of Jupiter unleashed her fury a second time. "The Lord cast down great stones from heaven upon them unto Azekah, and they died." The terrestrial axis tilted. Once more, the Earth quaked fiercely. Cities burned and fell to the ground. Above Beth-horon, the Sun stood still for hours. On the other side of the Earth, chroniclers recorded a prolonged night, lit only by the burning landscape. This occurred, Mexican records report, about fifty years after an earlier destruction.

As in its first encounter with the young comet, the Earth's surface was torn with great rifts and clefts, and hurricanes scoured the land. Strata pressed against strata, rising thunderously into mountains or engulfing cities. But the Earth and some of its inhabitants survived.

Anticipating renewed devastation following another fifty-year period, nations bowed down before the great fire goddess. With bloody orgies and incantations, they enjoined the dreaded queen of the heavens to remain far removed from the human abode. "How long wilt thou tarry, O lady of heaven and earth?" inquired the Babylonians. "We sacrifice unto Tistrya," declared a priest of Iran, "the bright and glorious star, whose rising is watched by the chiefs of deep understanding."

MARS

In both hemispheres, men fixed their gaze anxiously on the comet as, for centuries, it continued its circuit, crossing the orbits of both Earth and Mars. Before the middle of the eighth century B.C., astrologers observed dramatic irregularities in its wanderings. Viewed from Babylonia, Venus rose, disappeared in the west for over nine months, then reappeared in the east. Dipping below the eastern horizon, it was not seen again for over two months, until it shone in the west. The following year, Venus vanished in the west for eleven days before reappearing in the east.

But this time it was Mars, not Earth, that endured a cosmic jolt. Passing by the smaller orb, Venus pulled Mars off its orbit, sending it on a path that endangered the Earth. A new agent of destruction was born in the unstable solar system.

This occurred in the days of Uzziah, King of Jerusalem. (Lucian, the *Bamboo Books* of China, The Hindu *Surya-Siddhanta,* the Aztec *Huitzilopochtli Epos,* the Indo-Iranian *Bundahish,* etc., describe the reordering of Mars's and Venus's orbits.) Aware of the baleful meaning of irregular celestial motions, the prophet Amos, echoed by other observers of the sky, warned of new cosmic upheavals. Events soon vindicated the pessimistic seers.

As Mars drew near, the Earth reeled on its hinges. West of Jerusalem, half a mountain split off and fell eastward; flaming seraphim leaped skyward. Men were tossed into streets filled with debris and mutilated bodies. Buildings crumbled, and the Earth opened up.

These cataclysms were associated with the founding of Rome (placed by Fabius Pictor at 747 B.C.) and with the death of Rome's legendary founder, Romulus. "Both the poles shook," Ovid relates, "and Atlas shifted the burden of the sky . . . . The sun vanished and rising clouds obscured the heaven . . . ." Mars, the Lord of war, became the national god of Rome.

Much smaller than Earth, Mars could not equal Venus in destructive power. But again the Earth altered its course around the Sun. The old calendar, with 360-days and 30-day months, became outdated. Emperors and kings directed their astrologers to develop a new calendar.

BATTLE OF THE GODS

Mars and Venus now competed for the allegiance of men. Tribes moved from their homelands, confronting new enemies while petitioning Mars or Venus for a swift victory. Cities and temples were dedicated to the two planetary gods who determined the fate of nations.

The era of conflict between Mars and Earth continued

until 687 (or possibly 686) B.C. Hebrew prophets after 747 B.C. cried apocalyptically of upheavals yet to come. Reminding the Israelites of their passage out of Egypt, they declared that once more the whole Earth would quake, the Moon turn to blood, the Sun darken, and the Earth be consumed in blood, fire, and pillars of smoke.

The catastrophe, as Mars hurtled past the Earth, came in the year 721 B.C., on the day Jerusalem's King Ahaz was buried. Under the influence of Mars's passage, the Earth's axis tilted, and the poles shifted. Earth's orbit swung wider, lengthening the year.

Israelites observed the Sun hastening by several hours to a premature setting. Thereafter, the solar disk made its way across the sky ten degrees farther to the south.

Seneca records that on the Argive plain, in Greece, the early sunset came amid great upheaval. The tyrant Thyestes beckoned the entire universe to dissolve. The Great Bear dipped below the horizon. In the days which followed, states Seneca, "The Zodiac, which, making passage through the sacred stars, crosses the zones obliquely, guide and sign-bearer for the slow-moving years, falling itself, shall see the fallen constellations."

Once a peaceful, barely noticed planet, but now the "king of battle," Mars was still not finished with his work of destruction. In 687 B.C., a powerful Assyrian army led by Sennacherib marched toward Judah. On the evening of March 23, the first night of the Hebrew Passover, when Sennacherib and his army camped close to Jerusalem, Mars made a last, fateful approach to the Earth. A great thunderbolt—a "blast from heaven"—charred the soldiers' bodies, leaving their garments intact. The dead numbered 185,000. Ashurbanipal, Sennacherib's grandson, later recalled "the perfect warrior" Mars, "the lord of the storm, who brings defeat."

The same night, March 23, 687 B.C., in China, the *Bamboo Books* reveal, a disturbance of the planets caused them to go "out of their courses. In the night, stars fell like rain. The Earth shook." Romans would celebrate the occasion: "The most important role in the (Roman) cult of Mars appears to be played by the festival of *Tubilustrium* on the twenty-third day of March."

The Sun retreated by several hours. In certain longitudes the solar disk, which had just risen, returned below the horizon. In others, the setting Sun retraced its course, rising in the sky. The Hebrews witnessed the prolonged night of Sennacherib's destruction.

The Sun's retreat, due to a ten degree tilt of the Earth's axis, corrected the axis shift of 721 B.C. "So the sun returned by ten degrees, by which degrees it was gone down," reads Isaiah 38:8.

From one continent to another, men, oppressed with terror, watched Mars battle Venus in the sky, speed fiercely toward the Earth bringing blasts of fire, retreat, and engage Venus once more. Perhaps the most startling literary account of this *theomachy*, or battle of gods, is contained in Homer's *Illiad.* (Velikovsky's revised chronology places Homer later than 747 B.C.) As the Greeks besieged Troy, Athena (Venus) "would utter her loud cry. And over against her spouted Ares [Mars], dread as a dark whirlwind . . . . All the roots of many-founded Ida were shaken, and all her peaks." The river "rushed with surging flood" and "The fair streams seethed and boiled."

Mars was thrown out of the ring; Venus emerged a tame planet pursuing a near-circular orbit between Mercury and Earth. Where once it ranged high to the zenith, now it became the morning and evening star, never retreating more than 48 degrees from the Sun. Isaiah, who had witnessed the planet's destructive power, sang of its disgrace: "How art thou fallen from heaven, O Lucifer, son of the morning! How art thou cut down to the ground, which didst weaken the nations! For thou hast said in thine heart, I will ascend into heaven, I will exalt my throne above the stars of God."

Appendix C
The Night Vision of Charles D. Evans

In August, 1894, an article appeared in *The Contributor*, (Vol. 15, No. 20, pp. 638-647; the forerunner of *The Improvement Era*), entitled "A Dream." In the article, an account was given of a vision or dream which Charles D. Evans, a Patriarch living in Springville, Utah, had been given of the last days.

A perusal of the account, in light of the information in this book, is very revealing. Another remarkable aspect of this vision is the description of the New Jerusalem at the dawn of the Millennium.

A Dream

While I lay pondering, in deep solitude, on the events of the present, my mind was drawn into a reverie such as I had never felt before. A strong solicitude for my imperilled country utterly excluded every other thought and raised my feelings to a point of intensity I did not think it possible to endure. While in this solemn, profound and painful reverie of mind, to my infinite surprise, a light appeared in my room, which seemed to be soft and silvery as that diffused from a northern star. At the moment of its appearance the acute feeling I had experienced instantly yielded to one of calm tranquility.

Although it may have been at the hour of midnight and the side of the globe whereon I was situated was excluded from the sunlight, yet all was light and bright and warm as an Italian landscape at noon; but the heat was softer or more subdued. As I gazed upward, I saw descending through my bedroom roof, with a gently gliding movement, a personage clothed in white apparel, whose countenance was smoothly serene, his features regular, and the flashes of his eye seemed to shoot forth scintillations, to use an earthly comparison, strongly resembling those reflected from a diamond under an intensely-illumined electric light, which dazzled but did not bewilder. Those large, deep, inscrutable eyes were presently fixed upon mine, when instantly placing his hands upon my forehead his touch produced an indescribable serenity and calmness, a calmness not born of earth, but at once peaceful, delightful and heavenly. My whole being was imbued with a joy unspeakable. All feelings of sorrow instantly vanished. Those lines and shadows which care and sorrow impress upon us were dispelled as a deep fog before a blazing sun. In the eyes of my heavenly visitor, for such he appeared to me, there was a sort of lofty pity and tenderness infinitely stronger than any such feeling I ever saw manifested in ordinary mortals. His very calm appeared like a vast ocean stillness, at once overpowering to every agitated emotion.

By some intuition, or instinct, I felt he had something to communicate to soothe my sorrows and allay my apprehensions. Whereon, addressing me, he said:

"Son, I perceive thou hast grave anxieties over the perilous state of thy country, that thy soul has felt deep sorrow for its future. I have therefore come to thy relief and to tell thee of the causes that have led to this peril. Hear me attentively. Seventy-one years ago, after an awful apostasy of centuries, in which all nations were shrouded in spiritual darkness, when the angels had withdrawn themselves, the voice of prophets hushed and the light of Urim and Thummim shone not, and the vision of the seers was closed, while heaven itself shed not a ray of gladness to lighten a dark world, when Babel ruled and Satan laughed, and church and priesthood had taken their upward flight, and the voice of nations, possessing the books of the Jewish prophets [the Bible] had ruled against vision and against Urim, against the further visits of angels, and against the doctrine of a church of apostles and prophets, thou knowest that then appeared a mighty angel with the solemn announcement of the hour of judgment, the burden of whose instructions pointed to dire calamities upon the present generation. This, therefore, is the cause of what thou seest and the end of the wicked hasteneth."

My vision now became extended in a marvelous manner, and the import of the past labors of the Elders was made plain to me. I saw multitudes fleeing to the place of safety in our mountain heights. The church was established in the wilderness. Simultaneously the nation had reached an unparalleled prosperity, wealth abounded, new territory was acquired, commerce extended, finance strengthened, confidence was maintained, and peoples abroad pointed to her as the model nation, the ideal of the past realized and perfected, the embodiment of the liberty sung by poets, and sought for by sages.

"But," continued the messenger, "thou beholdest a change. Confidence is lost. Wealth is arrayed against labor,

labor against wealth, yet the land abounds with plenty for food and raiment, and silver and gold are in abundance. Thou seest also that letters written by a Jew have wrought great confusion in the finances of the nation which, together with the policy of many wealthy ones, has produced distress and do presage further sorrow."

Factions now sprang up as if by magic; capital had entrenched itself against labor throughout the land; labor was organized against capital. The voice of the wise sought to tranquilize these two powerful factors in vain. Excited multitudes ran wildly about; strikes increased; lawlessness sought the face of regular government. At this juncture I saw a banner floating in air whereon was written the words, "Bankruptcy, Famine, Floods, Fire, Cyclones, Blood, Plague." Mad with rage men and women rushed upon each other. Blood flowed down the streets of cities like water. The demon of bloody hate had enthroned itself on the citadel of reason; the thirst for blood was intenser than that of the parched tongue for water. Thousands of bodies lay untombed in the streets. Men and women fell dead from the terror inspired by fear. Rest was but the precurser of the bloody work of the morrow. All around lay the mournfulness of a past in ruins. Monuments erected to perpetuate the names of the noble and brave were ruthlessly destroyed by combustibles. A voice now sounded aloud these words: "Yet once again I shake not the earth only, but also heaven. And this word yet once again signifies the removing of things that are shaken as of things that are made; that those things that cannot be shaken may remain."

Earthquakes rent the earth in vast chasms, which engulfed multitudes; terrible groanings and wailings filled the air; the shrieks of the suffering were indescribably awful. Water wildly rushed in from the tumultuous ocean whose very roaring under the mad rage of the fierce cyclone, was unendurable to the ear. Cities were swept away in an instant, missiles were hurled through the atmosphere at a terrible velocity and people were carried upward only to descend an unrecognized mass. Islands appeared where ocean waves once tossed the gigantic steamer. In other parts voluminous flames, emanating from vast fires, rolled with fearful velocity destroying life and property in their destructive course. The seal of the dread menace of despair was stamped on every human visage; men fell exhausted, appalled and trembling. Every element of nature was a demon of wrathful fury. Dense clouds, blacker than midnight darkness, whose thunders reverberated with intonations which shook the earth, obscured the sunlight. Darkness reigned, unrivalled and supreme.

Again the light shone, revealing an atmosphere tinged with a leaden hue which was the precursor of an unparalleled plague whose first symptoms were recognized by a purple spot which appeared on the cheek, or on the back of the hand, and which, invariably enlarged until it spread over the entire surface of the body, producing certain death. Mothers, on sight of it, cast away their children as if they were poisonous reptiles. This plague, in grown persons, rotted the eyes in their sockets and consumed the tongue as would a powerful acid or an intense heat. Wicked men, suffering under its writhing agonies, cursed God and died, as they stood on their feet, and the birds of prey feasted on their carcasses.

I saw in my dream the messenger again appear with a vial in his right hand, who addressing me, said: "Thou knowest somewhat of the chemistry taught in the schools of human learning, behold now a chemistry sufficiently powerful to change the waters of the sea."

He then poured out his vial upon the sea, and it became putrid as the blood of a dead man, and every living soul therein died. Other plagues followed I forbear to record.

A foreign power had invaded the nation which, from every human indication, it appeared would seize the government and supplant it with monarchy. I stood trembling at the aspect, when, lo, a power arose in the west which declared itself in favor of the Constitution in its original form; to this suddenly rising power every lover of constitutional rights and liberties throughout the nation gave hearty support. The struggle was fiercely contested, but the Stars and Stripes floated in the breeze, and bidding defiance to all opposition, waved proudly over the land. Among the many banners I saw, was one inscribed thus: "The government based on the Constitution, now and forever;" on another "Liberty of Conscience, social, religious and political."

The light of the gospel which had but dimly shone because of abomination, now burst forth with a luster that filled the earth. Cities appeared in every direction, one of which, in the center of the continent, was an embodiment of architectural science after the pattern of eternal perfections, whose towers glittered with a radiance emanating from the sparkling of emeralds, rubies, diamonds and other precious stones set in a canopy of gold and so elaborately and skillfully arranged as to shed forth a brilliancy which dazzled and enchanted the eye, excited admiration and developed a taste for the beautiful beyond anything man had ever conceived. Fountains of crystal water shot upward their transparent jets which in the brilliant sunshine, formed ten thousand rainbow tints at once delightful to the eye. Gardens, the perfections of whose arrangement confound all our present attempts or genius, were bedecked with flowers of varied hue to develop and refine the taste, and strengthen a love for these nature's chastest adornments. Schools and universities were erected to which all had access; in the latter, Urims were placed for the study of the past, present and future, and for obtaining a knowledge of the heavenly bodies and of the construction of worlds and universes. The inherent properties of matter, its arrangements, laws, [and] mutual relations were revealed and taught and made plain as the primer lesson of a child. The conflicting theories of geologists regarding the formation and age of the earth were settled forever. All learning was based on eternal certainty. Angels brought forth the treasures of heaven which had lain hid in the womb of the dumb and distant past.

The appliances for making learning easy surpassed all conjecture. Chemistry was rendered extremely simple by the power which the Urims conferred on man of looking into and through the elements of every kind; a stone furnished no more obstruction to human vision than the air itself. Not only were the elements and all their changes and transformations plainly understood but the construction, operations, and laws of mind were thus rendered equally plain as those which governed the coarser elements. While looking through the Urim and Thummim I was amazed at a transformation which even now is to me marvelous beyond description, clearly showing the manner in which particles composing the inorganic kingdom of nature are conducted upward to become a part of organic forms; another astounding revelation was a view clearly shown me of the entire circulation of the blood both in man and animals. After seeing these things and gazing once more upon the beautiful city, the following passage of Scripture sounded in my ears: "Out of Zion the perfection of beauty, God shineth."

On this I awoke to find all a dream.

I have written the foregoing which is founded on true principles under the caption of a dream, partly to instruct and partly to check the folly of reading silly novels now so prevalent [in our society].

Appendix D

The following is a comparison of the Book of Exodus and the *Papyrus Ipuwer* account. It is included for the purpose of showing the remarkable similarity between the two accounts. Both are apparently eye-witness accounts of the events seen to occur in Egypt during the near collision of the Earth and the protoplanet Venus in approximately 1450 B.C.

Exodus	Papyrus Ipuwer
. . . there was blood throughout all the land of Egypt. (7:21.)	Plague is throughout the land. Blood is everywhere. (2:5-6.)
. . . all the waters that were in the river were turned to blood. (7:20.)	The river is blood. (2:10.)
And all the Egyptians digged round about the river for water to drink; for they could not drink of the water of the river. (7:24.)	Men shrink from tasting . . . and thirst after water. (2:10.)

. . . and the fire ran along upon the ground. (9:23.)

Forsooth, gates, columns, and walls are consumed by fire. (2:10.)

. . . and there was thick darkness in all the land of Egypt. (10:22.)

The land is not light. (9:11.)

And it came to pass, that at midnight the Lord smote all the firstborn in the land of Egypt, from the firstborn of Pharaoh that sat on his throne unto the firstborn of the captive that was in the dungeon. (12:29.)

Forsooth, the children of princes are dashed against the walls. (5:3, 6.) Forsooth, the children of princes are cast out in the streets. (6:12.) The residence is overturned in a minute. (7:4.) The prison is ruined. (6:3.) The towns are destroyed. (2:11.)

. . . and the hail smote every herb of the field, and brake every tree of the field. (9:25.)

Trees are destroyed. (4:14.) No fruit or herbs are found . . . (6:1.) Forsooth, that has perished which yesterday was seen. (5:12.) Upper Egypt has become dry (wastes?). (2:11.)

. . . the hand of the Lord is upon thy cattle which is in the field . . . there shall be a very grievous murrain. (9:3.)

All animals, their hearts weep. Cattle moan . . . (5:5.)

. . . gather thy cattle, and all that thou hast in the field. (9:19.)

Behold, cattle are left to stray, and there is none to gather them together. Each man fetches for himself those that are branded with his name. (9:2-3.)

. . . there was not a house where there was not one dead. (12:13.)

. . . there was a great cry in Egypt. (12:30.)

He who places his brother in the ground is everywhere. (2:13.)

It is groaning that is throughout the land, mingled with lamentations. (3:14.)

(*Kronos*, Vol. I, No. 2, pp. 17-22.)

Egyptological scholarship has vindicated Velikovsky's conclusions regarding the chronological placement of the Papyrus Ipuwer and its physical cataclysmic content. For more information see J. Van Seters, *Journal of Egyptian Archaeology*, 50, 1964, pp. 13-23; R. O. Faulkner, *Journal of Egyptian Archaeology*, 51, 1965, pp. 53-62; W. F. Albright, *Bulletin of the American Schools of Oriental Research*, 179 (1965), pp. 41-42; J. Van Seters, *The Hyksos* (New Haven, 1966), p. 120; also see L. M. Greenberg, "The Papyrus Ipuwer," *Pensee*, III (Winter, 1973), pp. 36-37.

Bibliography

Brooks, Melvin R., *L.D.S. Reference Encyclopedia,* Salt Lake City, Utah: Bookcraft, 1960.

Crowther, Duane S., *Prophecy, Key to the Future,* Salt Lake City, Utah: Bookcraft, 1962.

Gardiner, A. H., *Admonitions of an Egyptian Sage* from a hieratic papyrus in Leiden, 1909.

Mace, Wandle, *Journal of Wandle Mace,* Special Collections, Brigham Young University Library.

Ransom, C. J., *The Age of Velikovsky,* Glassboro, New Jersey: Kronos Press, 1976.

Smith, William, *The New Smith's Bible Dictionary,* Garden City, New York: Doubleday & Company, 1979.

Snow, Eliza R., *Sacred Hymns and Spiritual Songs,* Salt Lake City, Utah: Deseret News, 1912.

Talbott, Stephen L., ed., *Velikovsky Reconsidered,* New York, New York: Warner Books, 1977.

Talmage, James E., *Jesus The Christ,* Salt Lake City, Utah: Deseret Book Company, 1961.

Velikovsky, Immanuel, *Ages in Chaos,* Garden City, New York: Doubleday & Company, 1952.

Velikovsky, Immanuel, *Earth in Upheaval,* Garden City, New York: Doubleday & Company, 1955.

Velikovsky, Immanuel, *Theses for the Reconstruction of Ancient History,* 1945.

Velikovsky, Immanuel, *Worlds in Collision,* Garden City, New York: Doubleday & Company, 1950.

The Book of Jasher, Salt Lake City, Utah: J. H. Parry & Company, 1887.